Industrial Measurement and Control

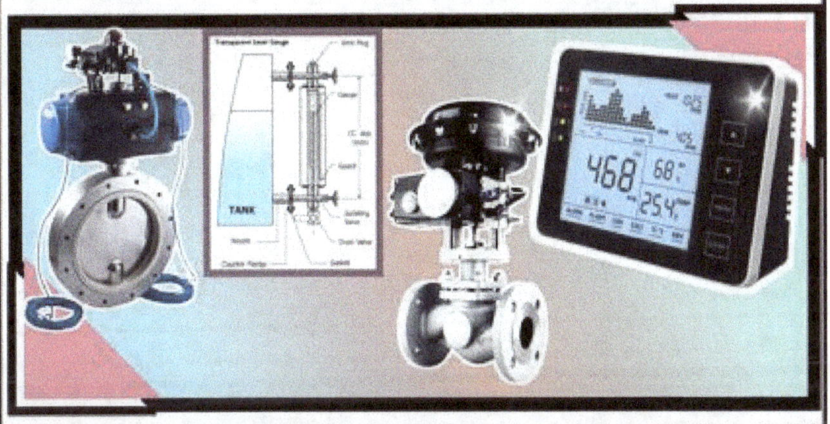

Zacchaeus A. Adetona

INDUSTRIAL MEASUREMENT AND CONTROL

Zacchaeus A. Adetona

Published 2024

ISBN: 978-978-787-134-8

*For enquiries and orders, please contact the author and the publisher at the indicated addresses.

Dedication

To my wife, Kemi, for her unflinching support for me and our family.

Acknowledgement

I want to express my sincere gratitude to Engr. Joel Ogunyemi (PhD) for his efforts in reviewing the manuscript of this book. My grateful thanks are also extended to Engr. Garba Abdulhamid for all his contribution in form of technical suggestions toward the completion of the textbook. I would also like to thank the entire staff of the Instrumentation and Control Option, Electrical Electronic Engineering Department, Federal Polytechnic, Ilaro for their technical and moral support in the course of preparation of the book. Thank you all.

Preface

As the industrial world continuously demonstrate great propensity towards automation, the role of instrumentation in the modern industries cannot be overemphasized. The electrical engineering students need to be well equipped for the subsequent engineering practice that they will encounter in the industry upon graduation. Thus, a concise and handy study material in the field of industrial instrumentation is needed. This engineering study aid Industrial Measurement and Control provides such highly needed information to help students in such departments as Electrical Electronic, Computer, Mechanical and System Engineering at the Higher National Diploma and all undergraduate levels. The fundamental knowledge of measurement, instrumentation and control has been covered in some previous texts. This book essentially touches on the theoretical background of industrial measurement, instrumentation, automation, control and the various devices employed to do these jobs. The modern methods and devices for industrial measurement and automation are covered. I have no doubt that this book will go a long way in helping the students in the affected departments grasp the much needed fundamental ideas and knowledge as demanded by the modern industries.

Foreword

Measurements and control are key aspects of industrial processes as quality products depend on them. It goes without saying that "whatever cannot be measured accurately cannot be controlled accurately as well". Control (or precise automatic control) is the heart of automation. Instrumentation is the science of measurement and it forms the basis of modern-day industrial automation. Industrial automation is, therefore, the use of measurement and control systems to perform tasks continuously, speedily, precisely and accurately as against manual production.

In this book, Industrial Measurement and Control, the author dealt extensively with the science and principles of industrial measurement, instrumentation, automation, and control devices. Drawing from his wealth of experience accumulated over long years in teaching and research; he has been able to package in one volume a book that covers many aspects of industrial measurements and control.

The book is structured into ten chapters for adequate coverage of knowledge in measurement and control. Chapters one and two deal with the fundamentals of industrial measurements. Chapter three is on transducers. Chapters four to seven extensively cover the main industrial variables which are temperature, pressure, flow and level. Installation of instruments was covered in chapter eight. Finally, chapter nine is on computer applications in industries.

The book will be a valuable asset to any student of electrical, electronics, mechanical and other related engineering in any tertiary institution who craves in-depth knowledge in the field. It can also serve as reference material for those practicing in the field as well including the hobbyists.

I therefore recommend this book for use by the students of tertiary institutions and others involved in the field of automation.

Engr. Dr. Joel Ogunyemi *(MNSE, MNIFEngM, FNIEE, COREN Registered)*

Head of Department
Department of Renewable Energy Engineering
Federal Polytechnic Ilaro
Ogun State.

Table of Contents

CHAPTER ONE

INDUSTRIAL MEASURING SYSTEMS

1.1   Introduction

In ages past, products and services were created by people using primitive methods including manual operation of molding, constructing, measuring, weighing, buying, selling etc. The advent of industrial revolution has brought about the development of various systems to carry out such tasks without employing the manual methods. The development of such systems requires the application of measurements and instrumentation. Measurement is born out of interaction of man with the physical world and is as old as civilization itself.

The industrial revolution that came about in the 1930's brought about the introduction of reliable instruments for measuring temperature. As the growth of manufacturing progressed, there came the need to have systems for continuous measuring of pressure, temperature, level, flow, etc. and the ever-present need for improvement of quality of products, and developments of new methods based on recently invented physical and chemical laws and effects. This called for the development of instrumentation systems that are very complex in many cases. Thus, measurement and instrumentation systems have become a very important body of knowledge that makes the world go around and exist the way we know it today. Measurements are carried out in manually or mainly using mechanical and electrical or electronic approaches. Measurement, instrumentation and automation are vital to industrial facilities and their products development. Industries rely heavily on such systems to design, implement, provide and maintain contemporary products and services.

The subject of measurement and control technologies is vast. Automation, field devices, and final control elements (valves) that measure and regulate different physical parameters (temperature, pressure, flow, etc.) are its main areas of focus. Its uses include the chemical, oil and gas, biological, and energy domains. Excellent uses of automation in buildings and airports are emerging, with instrumentation playing a major part. Another fascinating area that calls for interdisciplinary expertise is robotics, which includes instrumentation and control.

1.2   Definition and Aim of Measurement

Comparing an unknown quantity to a value that is presumed to be known is the process of measuring an unknown quantity. Both the physical universe and human existence depend on measurement. It provides us with an accurate and repeatable method for measuring the world we live in. In essence, measuring is the process of comparing an unknown value to a standard value that is known. The term "standard" refers to the latter. A system of measuring, also known as an analyzer, meter, scale, instrument, or instrument, is a tool used to make this comparison easier. Thus, comparing an unknown quantity with one of its standard values yields a measurement. Every measurement yields a unique number that is made up of a number and a unit of measurement with a name of its own. The result of each measurement is a specific number, consisting of a unit of measurement with its own name and a number indicating how often this respective unit is included in the measurement.

1.3   Importance of Measurement in the Industries

The main purpose of **measuring** in process industries and industrial production is to support the economics of industrial operations by improving product quality and production efficiency and to maintain proper operation. Measurements are of vital importance for these processes.

Good product quality is ensured in industries by measuring to the same standards throughout the process, from material receipt to processing, assembly, inspection, and shipping. Only a single inaccurate measurement can negatively affect the quality of the product. As a result, if defective products are mixed with good products, it leads to a decrease in yield. Further, defective products mixed with finished products lead to customers' dissatisfaction. Thus, it is critical that all measurements are accurate at every stage of the manufacturing process. The goal of accurate measurement is to guarantee that all production personnel are capable of weighing and that measuring apparatus is utilised and maintained correctly. These foundational ideas constitute a major concept of quality control known as "measurement control." The measurement equipment must be fit-to-function, or tailored to the application and prepared for usage even in challenging industrial settings, regardless of the actual application in production.

When measurements are carried out with accuracy, manufacturers save time and money, and improve the quality of their products. Industrial measurement engineers ask such questions as the following:

- Are the measuring devices working correctly?
- Are the measurement results accurate enough?
- How critical are the measurements?
- Will a wrong result cause economic losses or even death?

For any infrastructure project to be successful, precise measurements are essential. A thorough comprehension of the advantages of precise measurements by the production engineer in this domain could aid companies in improving and increase the market acceptability of their products.

Many industrial setups involve a large number of processes, thus precise instrumentation is necessary to ensure that everything runs properly. Process parameters including pressure, fluid level, temperature, and more must be understood by operators. Injuries to employees may result from inaccurate readings. Electrical instrumentation is primarily used in industrial settings to monitor process variables (such as temperature, pressure, and volume) and to manage equipment. These readings are used to automate controls that improve product quality and efficiency. Thus, industrial instrumentation is applicable for all of the following activities:

- Safety of life and property
- Better quality of products
- More economical production operations
- Longevity of equipment and machinery in operation.

1.4   Purpose of instrumentation Systems

There are two main purposes of industrial instrumentation namely primary and secondary purposes.

1.4.1   Primary purpose of instruments

The main function of an instrument involves measuring a process state, also known as a process variable or process parameter, such as pressure, temperature, level, flow, humidity, strain, displacement, or any of the many other potential process variables. Following measurement, the instrument's data might be used for further, or secondary, applications. A conversion process is employed within the instrument to change the measured variable such as temperature, pressure, flow, or chemical composition, into a more useful quantity, such as pressure, force, or potential difference.

The primary functions of instruments are carried out by various types of devices as transmitting, signaling, recording and indicating instruments. Transmitting devices send data about the measured amount to a distant location over a certain distance. The value of the measured quantity may never be made known, because it may be used for some other purpose. Signaling devices only display the measured quantity's general value or a range of values. A written record of the value of the measured quantity against some other variable or against time is provided usually on paper by recording instruments. For indicating instruments, the values of the quantity measured may be read on the scale to any fraction within the limitations of the instrument with the aid of some kind of calibrated scale and pointer or a digital readout.

1.4.2   Secondary purpose of instruments

The secondary purpose of instruments comprises the following:

i.   Controlling processes

Processes can be controlled manually or automatically. Both a flow meter and valve combination and a level-by-level meter and valve combination can be used to regulate liquid flow.

ii.   Process costing

Energy consumption records such as electricity and gas bills can be tracked with a household power meter (watt-hour meter) or gas meter. Similarly, meters are used at gas stations to assess process costs (the cost of fuel entering a vehicle's tank).

iii.   Maintaining personnel safety

Plants use full-body counter foils, dosimeters, etc. for personal protection. Radioactivity control, reactor start-up in case of high reactivity, and the use of AT in protection/control systems are examples of maintaining personal safety and plant safety.

iv.   Plant safety and efficiency

Various alarms and trips for reactors in abnormal conditions are available to maintain plant safety and efficiency.

v.   Redesigning process

Information from measurements can be used to redesign industrial processes. It is possible to get rid of flaws and weaknesses. By getting rid of unnecessary signals, process line piping networks can be made simpler.

1.5   Instrumentation and Automation

The versatile industrial automation involves the application of control networks and systems such as sensors, processors, actuators, and other machines to accomplish activities aimed at automating production. The rapidly increasing competition in the industry requires the provision of high-quality products at competitive prices. Industrial automation is currently the most powerful way to solve this problem. Industrial automation improves product quality and reduces production and construction costs by integrating innovative technologies and services. Different sectors are looking at different products to meet the challenges of automated devices.

Initial industrial automation started with basic conveyor belts transporting components on assembly lines. The term "automation" was first coined by engineers at Ford Motor Company, a pioneer in industrial automation and assembly line manufacturing. In the early days of the Industrial Revolution, production relied on human hands. But now we rely only on automated machines that are programmed to perform specific tasks flawlessly when given instructions. Basic activities were completed by machines, saving human labour. A vast array of devices, actuators, sensors, processors, and networks linking industrial surroundings make up modern industrial automation. The goal of contemporary industrial automation is to maximize the use of technology, from PLCs to artificial intelligence (AI), machine learning, and Internet of Things (IoT) gadgets. The following introduces the types and benefits of industrial automation that lead the world's industries, expand markets, and develop competition.

1.5.1   Benefits of industrial automation

Industrial automation has several advantages that manufacturers, OEMs, and industrial companies can leverage to improve operational efficiency. The advantages of modern industrial automation include:

- Boosting productivity
- Improving occupational safety
- Better decision making
- Improvement of quality
- Efficient use of resources
- Generation of higher profit

1.5.1.1 Increase productivity

As a result of spending less time on manual labor in an automated setting, personnel have more time, resources, and flexibility to concentrate on strategy, innovation, and technology. This translates into a notable rise in worker productivity. Production is slowed down by bottlenecks, performance problems, and downtime. When automated equipment and monitoring are connected via a single platform, issues like these can be identified early on and fixed before they have a negative impact on output. Data that enables predictive maintenance can reduce or eliminate downtime and performance issues. Real-time information ensures that the root cause of problems is identified, inventory adjusted, and production speeds monitored so that adjustments are made before bottlenecks occur.

1.5.1.2 Improve occupational safety

Firstly, industrial automation relieves workers of labour-intensive, hazardous, and time-consuming jobs. Industrial automation can make workplaces safer by lowering injuries brought on by moving heavy objects and doing repetitive tasks. This first advantage enhances daily operations by enabling staff members to concentrate on more difficult assignments.

1.5.1.3 Better decision making

When more devices are linked to and managed by industrial automation technologies, managers may build more precise models to identify new revenue streams, offer better bids, and interact with stakeholders. Connections can be made stronger. Industrial automation technology improves the capacity to track assets in the field, monitor and manage various sites remotely to expedite supply chain management, and give data for better decision making.

1.5.1.4 Higher quality of products

Automation equipment that is connected to the manufacturing line may gather data from almost any location, improve repeatability and fidelity, and notify anybody who has access. Early detection of quality changes can boost output, cut down on waste, cut down on rework, and boost revenue. Automated tools that can confirm and check quality also improve transparency and guarantee that produced items fulfill customer specifications before they are released from the manufacturer.

1.5.1.5 Efficient use of resources and higher profits

Faster throughput and overall increased manufacturing efficiency result from reduced machine transfer times in automated industrial processes. With manual labour, humans are prone to making mistakes. On the other hand, process automation tools are coded to ensure maximum accuracy and quality as well as minimizing errors. With automation, errors are reduced, and that can help reduce overall costs in a significant way. By automating activities, organisations can improve performance by reducing errors, improving quality and speed, and in some cases achieving results beyond human capabilities. Since automation shortens the time required to perform specific operations and decreases the number of workers required to execute various operational and safety tasks, labor costs will also be decreased. As a result, the unit cost per item will decrease overall, boosting earnings and creating new investment opportunities.

1.6   Types of Automation

Conveyor belts marked the beginning of industrial automation, which has progressed over time to include sophisticated AI and machine learning systems. It's critical to comprehend the various forms of industrial automation in order to pinpoint the precise applications where each technology excels. Various trends and solutions fit various applications, set-ups, and objectives.

Surveillance devices that are quick and simple to integrate into any kind of automation are the trend driving industrial automation. Data from real-time monitoring can be used to manage the workforce, modify the supply chain, and make decisions about production. Determining where and how industrial automation yields the best benefits is made easier for decision makers by monitoring. The industrial automation kinds are as follows.

1.6.1   Fixed automation

Hard automation, sometimes known as fixed automation, is characterized by high entry barriers, continuous processes, mass manufacturing, and predetermined duties. Seldom does this kind of industrial automation alter. High efforts are made to modify production or develop new items. More benefits come from programmable industrial automation for low volumes or short product life cycles.

1.6.2   Programmable automation

Tens to thousands of items can be manufactured with programmable automation, which is frequently used in batch manufacturing. Lead times and lot sizes are adjusted to accommodate for the increased number of changeovers that result from decreased production. Nonetheless, a lot of businesses are eager to boost output and increase uptime. Towards these ends, automation becomes increasingly flexible.

1.6.3   Flexible automation

The range of items that can be produced on a single machine or production line increases as industrial automation progresses since operations now require less human involvement and downtime. Precise electromechanical control is frequently a part of flexible automation. A CNC machine is a nice illustration. Downtime is decreased via automatic setup modifications. While this is effective for batch production, it also permits more customisation and high demand production.

1.6.4 Integrated automation

Processes, production lines, and other systems are connected through the integration of equipment and devices into a single control system. An approach to production that is more comprehensive is integrated automation. Industry 4.0 and the IoT have made it possible for autonomous production lines and equipment to interact over networks, enhancing their flexibility and enabling them to shift toward more customized and high-demand manufacturing.

1.7 Instrumentation and Automation

Specifically, instrumentation is the art of developing, building, and maintaining the measurement and control devices and systems that power manufacturing and research facilities critical to today's expanding economy. Industrial automation instrumentation offers many advantages, including: reduced labour costs by automating manual processes, improved traceability with sensors and computerized tracking.

Two fundamental aspects of modern industry are automation and instrumentation. The two contribute to making modern industry efficient and qualitative. While automation is the technique through which any process (a specific step in the production) is made automatic, instrumentation is the means by which the process is made automatic. Instrumentation is the means by which the process is made automatic and Automation is a term for technology applications where human input is minimized.

Generally speaking, when electronic devices and computer controls are utilised to control processes, automation is achieved. The goal of automation is to improve efficiency and reliability. However, in most cases, automation takes the place of laboratories. Here we give an example of the Continuous stirred-tank reactor (CSTR). In the reactor, the temperature has to be maintained at a particular level.

This can be carried out manually, but modern technology provides for an automatic alternative. The temperature of the reactor can be recorded at different points by means of a thermocouple. Then the average can be fed to a controller which checks it against the desired value (set point) and gives a control signal to the heating mechanism to either stop or increase heating.

This change in the manual process is called **automation** while the extra apparatus/ instruments used to reach the desired automation is termed **instrumentation**.

Instrumentation can be used for other purposes than automation such as:

1. Measuring of the final output of product for sale.

2. Monitoring a specific system from a remote location etc.

The results of application of automation have been tremendous in reducing errors and increasing both quality and quantity of products.

Automation engineers are responsible for ensuring that the entire manufacturing process runs automatically. Instrumentation and controls engineers pay special attention to the devices that perform control (controllers, PLCs, control valves, etc.) and the equipment that provides the necessary feedback to the controllers.

The two roles can mean the same thing or lead to the same work environment. However, whether they are exactly the same really depends on several factors. For example, in some companies an instrument engineer is someone who only deals with field instruments, while in other organisations the same job title is intended to cover control and safety systems as well.

Exercise

1. Explain the term "measurement" in engineering practice.
2. Explain how measurement is related to human existence.
3. Explain briefly the importance of measurement to the industries.
4. Enumerate and explain the following purpose of instrumentation system:
   a. primary
   b. secondary.
5. a. Explain the term "industrial automation".
   b. Explain the four types of automation systems.

6. What are the benefits of industrial automation?

7. Explain the duties of an automation engineer.

CHAPTER TWO

INSTRUMENTATION AND INDUSTRIES

2.1   Introduction

Instrumentation as a branch of engineering is widely employed in many manufacturing, production and several other industrial processes. These industries require a great deal of the application of instrumentation and automation. A high level of physical and applied science and engineering is required to put in place many of these industries. The systems involved in such industrial environments are developed by electrical, mechanical, instrumentation and computer engineers and scientists.

2.2   Instrumentation Industries

Industries where the knowledge of industrial instrumentation is applicable include any one of the following:

- Cement production
- Automobile production
- Aerospace industry
- Physical dimension(s) of an object
- Electronics manufacturing services/electronic industry
- Pharmaceutical drugs
- Power plants
- Oil and gas
- Automation industry
- Space industry
- Robotics
- Refineries
- Petrochemical

- Wastewater Treatment
- Food and beverage
- Several other kinds of manufacturing and production.

2.3 Industrial Measurement Quantities

Measurement plays a crucial role in industries which invariably makes it an indispensable aspect in modern industrial processes. From raw material quantification, sensing of process parameter to monitoring and packaging of industrial products and services, instruments are widely employed to achieve various results in production and service delivery.

Industrial measurement quantities include but not limited to the following:

a. Pressure and its variations
b. Volume of fluid stored in a vessel
c. Chemical concentration
d. Position, motion, or acceleration of a machine
e. Temperature
f. Level
g. Flow rate
h. Density
i. Viscosity
j. Humidity
k. pH
l. Force
m. Torque
n. Radiant energy
o. Thickness
p. Vibration
q. Particle counting
r. Radiation
s. Sound vibration

t. Velocity
u. Heat flow
v. Acceleration, etc.

The listed parameters are very important in measurement industries because they are measured from time to time to monitor and understand the operation of equipment, machinery and processes.

2.4   Significance of instruments in industries

By using an instrument, man can measure unknown quantities or variables that his unaided senses are unable to measure, so acting as an extension of human faculties. The following are other features of instruments in industries.

- A physical characteristic (such as velocity, pressure, flow rate, temperature, humidity, level, displacement, density, viscosity, etc.) would be sensed by the instrument, which would then process and translate it into a format and range that the observer could understand.
- The controls that allow the user to access, modify, and reply to the data should also be included in the instrument.
- An instrument is made up of a single unit that, when applied to an unknown variable (measurand), produces an output signal or reading. On the other hand, in more intricate measurement scenarios, a measuring device might comprise transducing components that transform the measurand into a comparable form. After undergoing some intermediary processing, the equivalent signal is supplied to the end devices so that the measurement results can be shown for record-keeping, control, and display.
- Man-made instruments are exact and sensitive in their response, and they also hold their qualities for a long time.

Exercise

1. Enumerate ten (10) instrumentation industries.
2. List ten (10) industrial measurement quantities.
3. Explain the significance of measurement in industries.
4. What are the significance of instruments to the industries.

CHAPTER THREE

TRANSDUCERS

3.1 Introduction

Engineering measurement and instrumentation often requires the conversion of a measurand from one form to the other. For example, to measure pressure requires two forms of conversion. An electrical signal is produced by first converting pressure to mechanical displacement. Most instrumentation systems have nonelectrical input quantities. To employ electrical methods and techniques for measurement, manipulation, or control, a nonelectrical quantity is first converted into an electrical signal by a transducer.

Transduction is the process by which energy is changed from one form to another. A transducer is, by definition, a device that, when activated by energy in one transmission system, provides energy to another transmission system in the same or a different form. Put another way, a transducer is an apparatus that transforms one type of physical phenomenon into another. Such physical phenomenon include heat, light, intensity, humidity, pressure, force, displacement, sound, magnetic flux, noise etc. LEDs, thermometers, microphones, and loudspeakers are a few common transducer examples. Examples of transducers in the industries include thermocouple, thermistor, strain gauge, optical encoder, bourdon tube, potentiometric and capacitive transducers.

Two important parts of a transducer are:

- Sensing element
- Transduction element

The sensing element is the part of a transducer that responds to the physical sensation. Thus, the response of the sensing element depends on the physical phenomenon. The transduction element of the transducer, also called the secondary transducer, converts the output of the sensing element into an electrical signal.

Input and output transducers are the two different types of transducers. Physical energy is taken in by an input transducer or sensor, which transforms it into a readable electrical signal. For instance, a thermocouple transforms thermal energy into an electrical signal that can be sent over wires. Conversely, an actuator, also known as an output transducer, receives electrical impulses and transforms them into other types of energy. An example is an electric motor that converts electricity into mechanical motion.

3.2   Electrical Transducers

Electrical transducers take various forms such as resistive, capacitive and inductive type transducers. An electrical transducer is a sensing device that converts a physical, mechanical, or optical (non-electrical) property that needs to be measured directly into an electrical signal (voltage, current, or frequency) via an appropriate mechanism.

Transducers can be categorized based on a variety of factors, including their intended use, energy conversion technique, output signal type, and more. Transducers are grouped based on the electrical principle they use in the following table. Passive transducers are those that need an external power source, like some of these transducers. These transducers cause variations in voltage or current, which can be used to monitor changes in electrical properties like capacitance, resistance, and so forth.

The self-generating kind of transducer is another group. When they are triggered by a physical energy source, they generate an analog voltage or current. They can function without the need for external electricity. The following table lists the various classes of transducers along with their typical uses and principles of operation.

Table: Types of Transducers

Passive Transducers		
Class of Transducer	Principle of Operation	Typical Application
1. Potentiometric device	Variation in resistance caused by the placement of a slide contact by an outside force in a potentiometer or bridge circuit.	Pressure, displacement, position.
2. Resistance strain gauge	Variation in a wire's resistance caused by compression or elongation as a result of external stress.	Force, torque, displacement.
3. Thermistor	Certain metal oxides with negative temperature coefficients of resistance exhibit temperature-dependent variations in resistance.	Temperature
4. Pirani gauge or hot wire meter	Variation in a heating element's resistance caused by a gas's steam convection cooling.	Gas flow, gas pressure
5. Resistant thermometer	Changes in a pure metal wire's resistance at a high positive temperature coefficient.	Temperature, radiant energy
6. Photoconductive cell	Changes in incident light resistance using a photoconductive cell as a circuit element.	Photosensitivity relay

Capacitance		
7. Variable capacitance pressure gauge	Distance between two parallel plates is varied by an external applied force.	Displacement pressure.
8. Capacitor microphone	The capacitance between a moving diaphragm and a fixed plate is varied by sound pressure.	Speech, music, noise
9. Dielectric gauge	Variation in capacitance by changes in the dielectric.	Liquid level, thickness
Inductance		
10. Magnetic circuit transducer	Ac-excited coil self-inductance or mutual inductance is varied by altering the magnetic circuit.	Pressure, displacement
11. Reluctance pickup`	Reluctance of a magnetic circuit is varied by changing the position of iron core of a coil.	Pressure, displacement, vibration, position
12. Differential transformer	By moving the magnetic core of a transformer through an externally applied force, the differential voltage between its two windings is varied.	Pressure, force, displacement, position
13. Eddy current gauge	A coil's inductance varies when an eddy current plate is nearby.	Displacement, thickness
14. Magnetostriction gauge	Stress and pressure vary the magnetic characteristics.	Force, pressure, sound
15. Voltage and current Hall effect pickup	A potential difference is produced across a germanium semiconductor plate when an applied	Magnetic flux, current

	current and magnetic flux interact.	
16. Ionization chamber	Ionization of gas by radioactive radiation causes an electron flow.	Particle counting, radiation
17. Photoemissive cell	Secondary emission on photoemissive surfaces as a result of incoming radiation.	Light and radiation
18. Photomultiplier tube	Secondary electron emission due to incoming radiation on photosensitive relay.	Light and radiation
Self-Generating Transducers		
Class of Transducer	**Principle of Operation**	**Typical Application**
19. Thermocouple and thermopile	When the junction of two dissimilar metals or semiconductors is heated, an electromagnetic field (emf) is produced across that junction.	Temperature, heat flow, radiation
20. Moving coil generator	A coil moving in a magnetic field generates voltage.	Velocity, vibration
21.Piezoelectric pickup	When an external force is applied to some crystalline materials, such as quartz, an electromagnetic field (emf) is produced.	Sound vibration, acceleration, pressure variations
22. Photovoltaic cell	Radiant energy stimulates a cell, which results in the generation of voltage in a semiconductor junction device.	Light meter, solar cell

3.3   Selecting a Transducer

The first and possibly most crucial step in getting an accurate result from a measurement system is choosing the right transducer. Prior to choosing a transducer, the following basic inquiries should be taken into account:

- What physical quantity needs to be measured?
- What transducer concept works best for measuring this quantity?
- What accuracy is required for this measurement?

  The kind and range of the measurand are taken into account when responding to the first query. Understanding the transducer's input and output characteristics—which must work with the recording or measurement system—is necessary to respond to the second question. Such factors include:

i.   Fundamental transducer parameters
ii.   Physical conditions
iii.   Ambient conditions
iv.   Environmental conditions
v.   Compatibility of the associated equipment

The choice of a transducer requires the knowledge and proper consideration of a number of the above factors.

3.4   Resistive Type Transducers

Numerous applications use resistive transducers to measure various parameters, including fluid velocity, load, temperature, pressure, mechanical strain, displacements, and force. These devices operate on the principle of resistance changing as a result of the measured quantity. Often referred to as variable resistance transducers or resistive sensors, these transducers are most commonly utilised in the calculation of many physical parameters, including force, displacement, temperature, vibration, pressure, and vibration. They are employed in primary and secondary sensing elements. Since the output of the primary transducer can be used as an input by the resistive transducer, they are typically employed as secondary sensing components. Its obtained output is calibrated based on the input quantity and yields the input value straight away. Potentiometers, resistive pressure transducers, thermistors, photoresistors, strain gauges, and light-dependent resistors are a few types of resistive transducers.

3.4.1   Potentiometer Transducers

In a potentiometer transducer the output is resistance which varies under the action of displacement which is the input quantity. One application for a resistive divider is as a potential divider or potentiometer. In other words, a potentiometer transducer is essentially a potentiometer with its wiper moved by the input quantity.

3.4.2   Linear Potentiometric Transducer

A linear potentiometric transducer consists of a resistance element wound on a former and a wiper. The input quantity is attached to the wiper and causes a displacement on it as the quantity varies. The resistance element is wound with an alloy of platinum – Iridium, constantan, Ni-chrome, or Fe-Cr-Al. The displacement caused on the wiper in turn produces a variation in the output resistance of the transducer. The figure below gives two examples of linear potentiometric transducer.

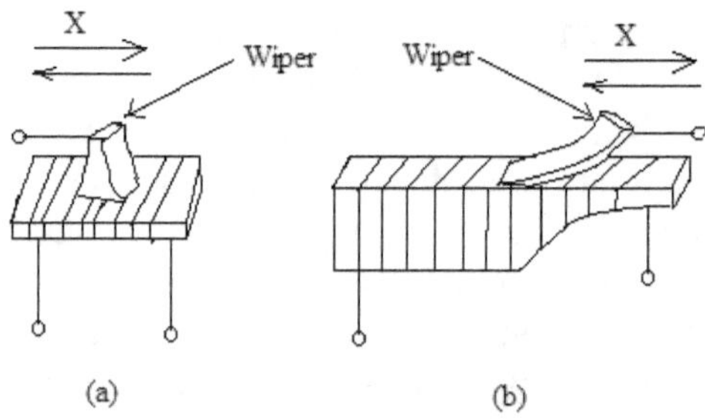

As can be seen from the figure, the wiper can be moved on the potentiometer both ways along X. As this takes place, a change in resistance across the transducer terminals results.

3.4.3 Angular Potentiometric Transducers

Angular potentiometric transducer type is similar to a linear type in that it also consists of a resistance element wound on a former on which a wiper is attached. The distinguishing feature of the angular potentiometer transducer is that the wiper moves in a circular way. That is, it is capable of making only angular displacement and not linear. This is demonstrated in the following Figure. The wiper is made of either a piece of wire or an assembly of flat springs. In both cases use is made of either a pure metal (platinum or silver) or an alloy (Platinum – Iridium or phosphor bronze). As the measured quantity varies, a corresponding angular displacement is caused on the wiper which in turn results in corresponding variation in the resistance across the terminals of the transducer.

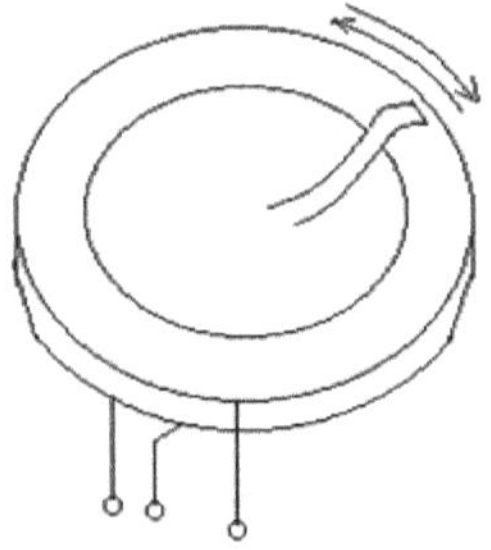

3.4.4   Transfer Function of a Potentiometric Transducer

Consider a potentiometric transducer show below. The source of emf is a constant voltage that drives a constant current i through the circuit.

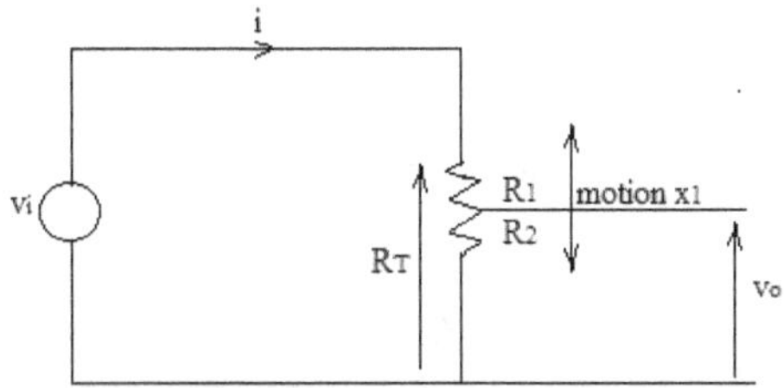

x_i = Input displacement in meters

x_t = maximum displacement possible

R_1 = |Resistance due to input displacement

R_2 = Resistance in the remaining part of the potentiometer not covered by the displaced wiper

R_T = Total resistance of the potentiometer

From the above sketch, we can deduce that:

$$R_2 = \frac{x_1}{x_T} R_T \qquad\qquad (i)$$

$$vi = i\,(R_1 + R_2) \qquad\qquad (ii)$$

$$v_o = iR_2 = i\frac{x_1}{x_T} R_T \qquad\qquad (iii)$$

But $\quad R_T = R_1 + R_2$

$$R_2 = R_T - R_1$$

$$v_i = i[R_1 + (R_T - R_1)]$$

$$v_i = iR_T \qquad\qquad (iv)$$

∴ Transfer function is given as:

$$\frac{vo}{vi} = i\frac{x_1}{x_t} R_T \,/\, iR_T$$

$$\frac{v_o}{v_i} = \frac{x_1}{x_t}$$

Using equations (ii) and (iv) we can express the transfer function as a ratio of resistance thus:

$$\frac{v_o}{v_i} = \frac{iR_2}{i(R_1+R_2)}$$

$$\frac{v_o}{v_i} = \frac{R_2}{R_T}.$$

Examples

1. In the figure, a displacement transducer with a 6 cm shaft stroke is used. The potentiometer has a total resistance of 5 kΩ and an applied voltage, V_T of 5 V. What is the output voltage V_0 when the wiper is 1.9 cm from B?

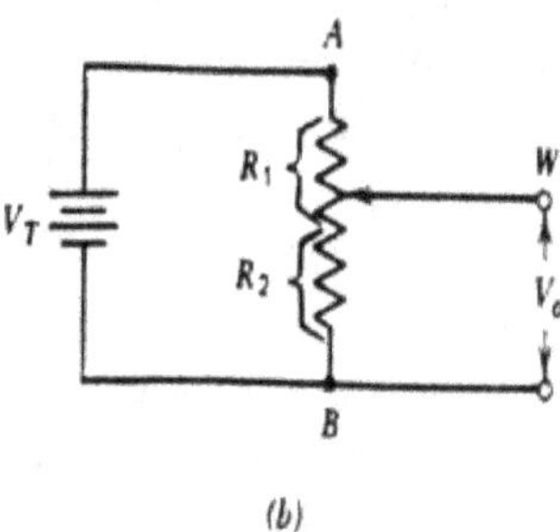

(b)

Solution

$$R_2 = \frac{x_1}{x_T} R_T$$

$$R_2 = \frac{1.9}{6} \; x \; 5000$$

$$R_2 = 1583.33 \; \Omega.$$

$$\frac{v_o}{v_i} = \frac{R_2}{R_T}$$

$$v_o = \frac{R_2}{R_T} v_i$$

$$= \frac{1583}{5000} \; x \; 5$$

$$v_o = 1.53 \; V.$$

3.5   Resistance Change Transducers

The potentiometric transducer earlier described is an example of resistance change transducers. It was mentioned that a displacement on the slider results in a change in resistance of the transducer. In a resistance change transducer, a change in the measurand typically results in motion or displacement on a sliding contact. This movement or displacement then causes a change in the circuit resistance and, ultimately, a change in current. Consider the following figure.

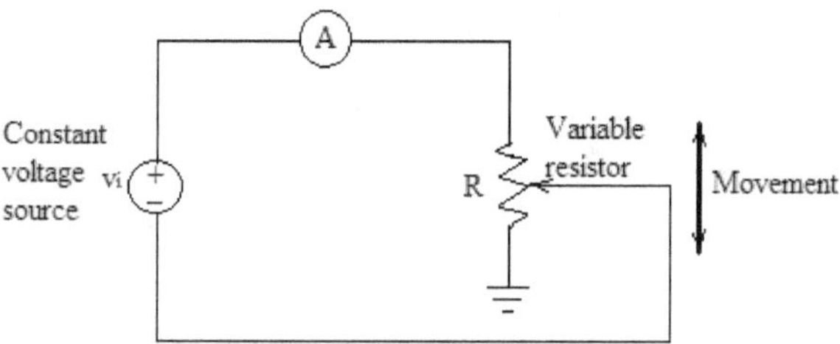

In order to ensure that the current I in the circuit is only reliant on variations in the resistance R, the voltage must remain constant in magnitude regardless of the circuit resistance in order for the transducer to operate satisfactorily. This sliding contact is connected to the motion or displacement that is being watched. A continuous display of the measurand would result from calibrating the ammeter in the appropriate units. The resistance change transducer has a few shortcomings, which include:

i.   When the measurand is in the zero or reference position, current will be flowing in the circuit if a plus and minus fluctuation is to be noted.

ii.   ii. The displacement per second action is restricted to a few cycles. About 3 x 106 cycles of operation are the typical lifespan of this kind of transducer.

3.6   Resistance Strain Gauge

A small, wafer-shaped instrument called a strain gauge is used to measure applied strain on a range of materials. It is an example of a passive transducer, which changes resistance in response to a mechanical displacement. Because it operates on the basis of mechanical strain applied to a piece of wire, the device is dubbed strain gauge.

Tensile force applied to an electrical wire will cause it to stretch, increasing its length by δL. The change in length, δL, is proportional to the load and the wire will return to its original length L when the load is removed, so long as the elastic limit is not exceeded. Mechanical strain E is thus defined as:

$$\epsilon = \frac{\delta L}{L} \qquad\qquad (i)$$

A decrease in the cross-sectional area A coincides with the change in the wire's length, δL, caused by the applied load. Given that a conductor's resistance R depends on its resistivity, cross-sectional area, and length, thus:

$$R = \frac{\rho L}{A} \qquad\qquad (ii)$$

The stretched wire will thus have more resistance as a result of both the increase in length and the decrease in area. Additionally, there is a little but significant change in the material's resistivity, which also contributes to the resistance change.

The strain gauge is typically set up by mounting a folded resistance wire on paper or Bakelite backing and placing it in a grid.

Poisson's Ratio and Strain Sensitivity of a Strain Gauge

A wire's electrical resistance varies with its length, as was previously mentioned. Let δR be the resistance change. Strain sensitivity is a property of the strain gauge that connects its change in length per unit length to its change in resistance per unit resistance. It is represented as k and defined as:

$$K = \frac{\delta R / R}{\delta L / l} \qquad (iii)$$

Using equation (i) we have that

$$K = \frac{\delta R / R}{E} \qquad (iv)$$

Where E is the strain in the lateral direction

From equation (ii)

$$R = \frac{\rho L}{A} = \frac{\rho L}{\left(\frac{\pi}{4}\right) d^2} \qquad (v)$$

Where d = diameter of the conductor.

When under tension, the resistance is affected by δL and δd as follows (where R_s = resistance of stretched wire)

$$R_s = \frac{\rho (l + \delta l)}{\left(\frac{\pi}{4}\right)(d - \sigma d)^2} = \frac{\rho L \left(1 + \frac{\sigma L}{L}\right)}{\left(\frac{\pi}{4}\right) d^2 \left(1 - \frac{\sigma d}{d}\right)} \qquad (vi)$$

A term, Poisson's ratio (μ) defined as the ratio of the strain in the lateral direction, can be used to simplify equation (vi) thus:

$$\mu = \frac{\sigma d / d}{\sigma L / L} \qquad (vii)$$

For most metals, μ in the range of 0.25 to 0.35.

$$Rs = \varphi \frac{i}{\left(\frac{\pi}{4}\right) d^2} \left(\frac{1 + \frac{\Delta L}{L}}{1 - \frac{2\mu \Delta L}{L}} \right) \qquad (viii)$$

Equation (viii) is of the form

$$Rs = R + \Delta R = R\left\{1 + (1 + 2\mu)\frac{\delta L}{L}\right. \qquad (ix)$$

$$k = \left.\frac{\frac{\delta R}{R}}{\frac{\delta L}{L}}\right. = 1 + 2\mu \qquad (x)$$

Example

Attached to a steel component under 1,050 kg/cm^3 of stress is a resistance strain gauge with a gauge factor of 2. About 2.1 x 10^6 kg/cm^2 is the modulus of elasticity of steel. Determine the strain gauge element's change in resistance as a result of the applied stress.

Solution

From Hookes law, $E = \dfrac{Stress}{Youngs\ modulus} = \dfrac{S}{\in}$

$$E = \frac{1050}{2.1X10^6} = 5X10^{-4}$$

$$k = \frac{\sigma R\,/\,R}{E}$$

$$\frac{\sigma R}{R} = KE$$

$$= 2\ x\ 5\ x\ 10^{-4} = 10^{-3}$$

$$= 0.1\%$$

3.7   Bonded and Unbonded Strain Gauges

Any strain gauge designed to be affixed to or bounded by a range of materials in order to measure applied strain is called a bonded strain gauge. Examples are metallic types of strain guage typical of which is constantan, a copper–nickel alloy consisting of 60 / 40 Cu / Ni.

Further examples of metallic strain gauges are made of nichrome (nickel–chrome), which is used to measure static strain up to 375°C; dynaloy (nikel–iron), which has a high gauge factor and a high resistance to fatigue; and platinum–tungsten alloy, which is used to measure dynamic strain up to 850°C and offers excellent stability and high resistance to fatigue at elevated temperatures.

An armature supported in the center of the stationary frame makes up an unbonded strain gauge. The armature can only move in one direction, which is restricted by four filaments of strain-sensitive wire wound between rigid insulators mounted on both the armature and the frame.

Various configurations of the unbonded strain-gauge transducer can be built based on the intended application. Its primary function is as a displacement transducer. Examples of such displacement transducers are diaphragm, bellows, bourdon tubes, straight tube, mass cantilever and pivot torque.

3.8   Capacitive Transducers

In a capacitive transducer, any change in the distance between the parallel plates, d, results in an equal change in the capacitance of the capacitor. This is given in the following relation:

$$C = \frac{kAEo}{d} \ (Farads)$$

where   A = the area of each plate (m^2)
         d = the plate spacing (m)
         $Eo = 9.85 \times 10^{-12}$ (F/m)
         K = dielectric constant

Consider a capacitive transducer shown in the following figure. Applying a force to the diaphragm makes it work like a single plate of a basic capacitor. As a result, the distance of separation between the diaphragm and the static plate changes.

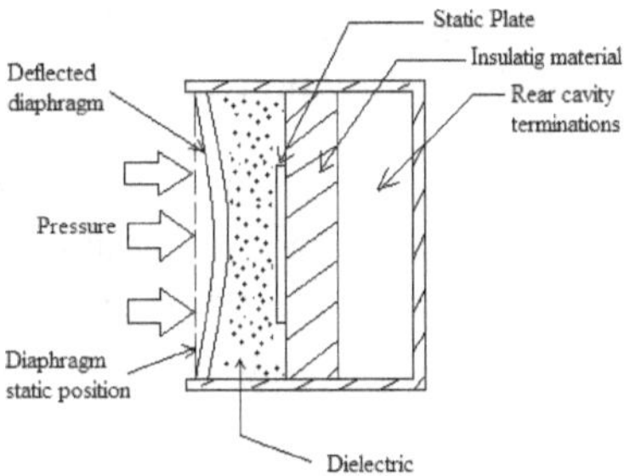

An AC oscillator circuit is typically used to measure the consequent change in capacitance, an AC bridge could also be used. The oscillator's frequency changes due to the transducer, which is a component of the oscillatory circuit. The applied force is measured by this change in frequency.

The capacitive transducer can assess both static and dynamic phenomena and has an excellent frequency response. Its shortcomings include its sensitivity to temperature radiations and the potential for irregular and distorted signals as a result of its lengthy head length.

3.9   Inductive transducer

In an inductive transducer, the inductance in a single coil or the change in the inductance ratio of two coils are both used to detect force. Each time, the force being measured causes the ferromagnetic armature to move, changing the magnetic circuit's outcome.

The way that magnetic flux links a coil's terms determines its inductance. It is possible to produce a change in inductance proportionate to the measurand by employing an appropriate measurement to relocate a converter and alter the magnetic flux connections in a coil. Either an oscillatory circuit's resonance frequency or the bridge balance's amplitude change can be used to quantify this change in inductance. The inductance of the flux path can be varied in one of the following two ways:

i)   Making an armature to change the flux path's resistance, allowing for either linear or angular displacement. This is depicted in the following figure. Measurements of pressure, acceleration, force, displacement, and position can be made with this method..

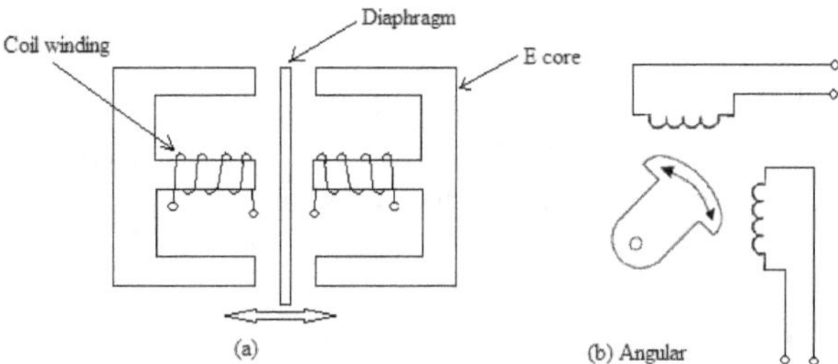

Variable reluctance transducer

ii)   Displacement of a high-magnetism slug of material that is confined to travel inside a center-tapped coil wound around a ferromagnetic core. The inductance of the two halves of the slug is equal while it is positioned centrally, as seen in the figure below. When the slug moves to one side, the inductance of one half increases and the other half drops.

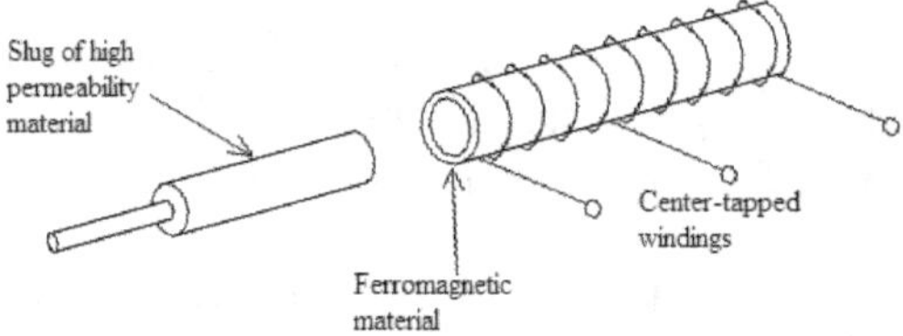

The inductive transducer offers a continuous resolution, a reasonably high output, and it responds to both static and dynamic measurements. Its disadvantages are:

i)   Limited frequency response caused by the construction of the force-summing member (i.e. diaphragm).

ii)   Its performance can be erratic due to the influence of external magnetic fields.

3.10   Temperature measurements

A highly pure platinum, copper, or nickel wire is used as the sensitive element in a resistance thermometer, which gives a precise resistance value at every temperature in its range. The following equation represents the relationship between conductor resistance and temperature

$$Rt = R_{ref} \left(1 + \alpha \Delta t\right)$$

where R_t = Resistance of the conductor as temperature t ($^\circ$c)

R_{ref} = Resistance at the reference temperature, usually 0°C

α = temperature coefficient of resistance

Δt = difference between operating and refernece temperature.

The resistance of almost all metallic conductors rises as temperature rises because they have a positive temperature coefficient of resistance, α. Because of their negative temperature coefficient of resistance, materials like germanium and carbon become less resistant as the temperature rises. In a temperature sensor element, a high value of α is preferred since it causes a significant change in resistance for a comparatively little temperature change. We shall discuss about some of the most popular temperature transducers presently.

3.11   Thermistor as Transducer

Thermistors, also known as thermal resistors, are semiconductor devices that exhibit the characteristics of resistors with a high temperature coefficient of resistance, typically negative. They are highly suitable for precise temperature monitoring, control, and compensation because of their high sensitivity to temperature changes. In these kinds of applications, they are frequently utilised, particularly at lower temperatures between -100°C to 300°C.

The thermistor is a clear choice for a temperature transducer due to its comparatively large resistance changes per degree change in temperature (a property known as resistivity). Any change in temperature causes the thermistor resistance, which in turn produces a change in the circuit current. The thermistor is connected in a straightforward series circuit with a battery and a microammeter. Measurements by the microammeter is calibrated in terms of temperature. A more sensitive arrangement would consist of a Wheatstone bridge arrangement shown below.

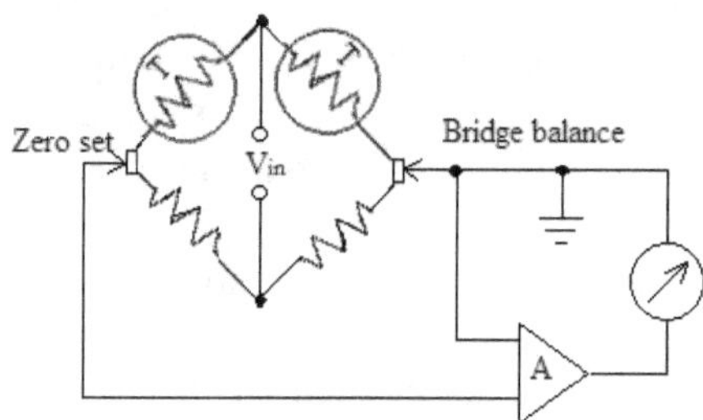

3.12   Thermocouple as Transducer

A thermocouple is made up of two different metal wires or bars that are connected at one end (referred to as the sensing or hot junction) and terminated at the other end (referred to as the reference or cold junction), which is kept at a constant temperature (referred to as the reference temperature). An emf is created when there is a temperature differential between the reference and sensing junctions, which results in a current flowing through the circuit. This is illustrated in the following diagram:

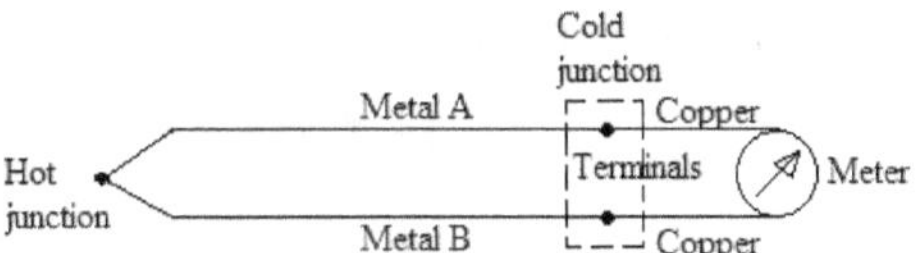

The meter indication will be proportionate to the temperature differential between the hot junction and the reference junction when the reference junction is terminated by a meter or recording device as demonstrated above. This is termed thermometric effect. The materials used to make the wire also affect the emf produced. The emf is given as

$$e = At + \frac{1}{2}Bt^2 + \frac{1}{3}Ct^3$$

Where e = emf produced by the couple

        t = sensing junction temperature

        A, B, C = constants of the thermocouple material

3.13   Laws of thermocouple

There are three laws guiding the production of thermal voltage in a thermocouple. These are:

3.13.1 Laws of homogenous material

This law asserts that no matter how the materials differ in the cross-section, heat alone cannot support the thermoelectric current in a circuit of a single homogenous material. Stated differently, as long as all of the wires on the thermocouple are composed of the same material, temperature variations in the wiring between a material's input and output have no effect on the output voltage.

3.13.2 Laws of intermediate material

According to this law, there is zero algebraic sum of the thermoelectric forces in a circuit made up of any number of different materials. In other words, if a third metal is placed into either wire and two junctions are at the same temperature, and all of the junctions are at the same temperature, the new metal produces no net voltage.

3.13.3 Laws of intermediate temperatures

The law of intermediate temperatures states that if two dissimilar homogenous materials are at temperature T1 and T2 when they produce thermal emf 1, and at temperature T2 and T3 when they produce thermal emf 2, then the emf generated at T1 and T3 will be (emf 1 + emf 2).

3.14 Variable Differential Transformer

This transducer calculates force as a function of the transformer's ferromagnetic core displacement. The following figure illustrates the fundamental design of a linear variable differential transformer (LVDT).

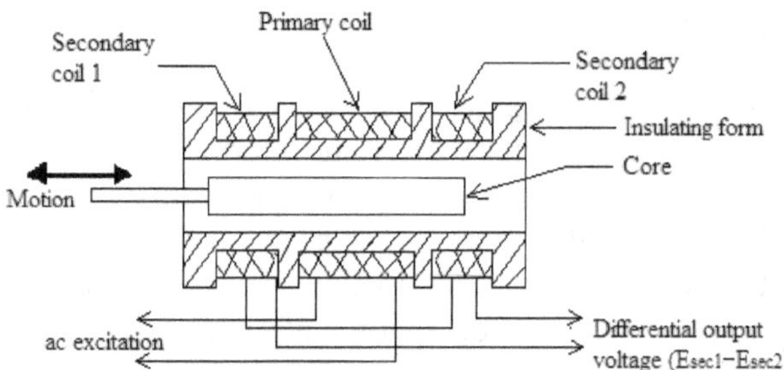

Rather than a variation of self-inductance, the change in inductance in this kind of transducer is a variation of mutual inductance between windings. As seen in the above figure, this variation is brought about by the movement of a ferromagnetic core. On an insulating, non-magnetic former, the coils are coiled. The LVDT's analogous circuit is displayed as follows. (Fig a).

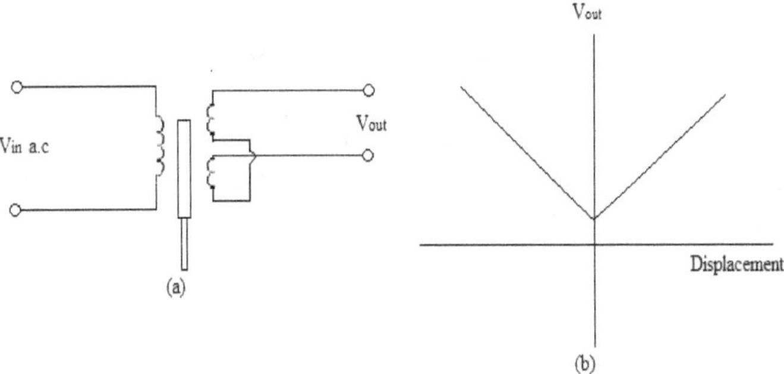

Fig (b) shows the voltage waveform which increases in magnitude for displacement either side of the centre or zero position.

To utilise the differential transformer transducer, a suitable "measurand to displacement" converter is placed between the quantity and the differential transformer. It measures total, steady, and dynamic displacements and is a robust, low impedance device with minimal to no friction.

3.15   Mechanical Transducers

A group of fundamental sensing components known as mechanical transducers react to variations in a physical quantity by producing a mechanical output. Because they translate mechanical outputs into physical values or vice versa, they are known as mechanical transducers. Because they provide mechanical output signals, mechanical transducers are distinguished from electrical transducers. Mechanical transducers can produce a wide range of outputs, including displacement, force (or torque), pressure, and strain. Mechanical transducers produce physical or mechanical signals at the output in response to any physical changes in the system. In a mechanical transducer, frictional losses take place. This is due to friction between moving parts. These losses lead to reduced efficiency and heat generation, which can have various effects on the transducer's performance and operation.

Mechanical transducers, which measure pressure, velocity, flow rate, and water force, include bellows, diaphragm, manometers, and bimetallic strips. Hydropneumatic transducers, on the other hand, include orifices, venturi, pitot tubes, vanes, and turbines. The operating principles of some of these transducers are discussed in chapter six.

Exercise

1a. What are electrical transducers?
b. Give two transducers for each of the following physical quantities:
(i) displacement (ii) temperature (iii) pressure (iv) light
c. Name four types of electrical pressure transducer and describe the operating principles and one application of each type.
d. Discuss three important factors you would consider when selecting a transducer for a measurement system.

2a. Discuss the principle of operation of a potentiometric transducer.
b. Describe how a linear and an angular potentiometric transducer work.
c. Derive the transfer function of a potentiometric transducer
d. What are the drawbacks of a resistance change transducer?

3a. What is a strain gauge? Describe its operating principle.
b. Distinguish between a bonded and an unbonded strain gauge giving two examples of each.

4a. Explain the term mechanical transducer.
b. List six examples of mechanical transducer.
c. Distinguish between electrical and mechanical transducer.

5a. Describe (i) a capacitive transducer (ii) an inductive transducer (iii) variable differential transformer, using appropriate diagrams.
b. Discuss with the aid of suitable diagrams the principle of operation of:
    i.   Thermocouple used as a transducer
    ii.   Thermistor used as a transducer.
c. Explain the three laws of thermocouple.

CHAPTER FOUR

TEMPERATURE MEASURING SYSTEMS

4.1   Introduction

The concept of temperature has a long history dating back to the beginning of human development. The methods of temperature measurements have changed rapidly in recent years and are continuing to do so. Over the years, the concept of temperature and its measurement have become parts of history of development of science. The concept of temperature has been the most difficult of common properties of matter to define in clear terms. A review of the historical development of temperature is beneficial as it was the development of earliest temperature measuring devices that led to the definition of temperature. Temperature monitoring is crucial in today's industry to preserve product quality, save associated costs, and minimize faults. Extreme temperatures have the potential to modify the chemical composition of products, resulting in changes to their texture or appearance. In the industrial production process, temperature is a potential source of inaccuracy. As a result, it is crucial to give it extra consideration in order to guarantee that the finished product is of the highest caliber. Furthermore, temperature monitoring and control systems preserve the health and safety of workers in the workplace as well as those of the general public. For example, in the chemical industry, improper temperature control can set off a series of events that can result in fire, explosion, the spread of poisonous gasses, and other catastrophic outcomes. All industries must continuously monitor and control the temperature of their facilities to prevent incidents like this one. In order to achieve this, temperature monitoring systems are installed in industrial facilities, enabling real-time temperature setting and control. The following are functions of industrial temperature measurement and control.

- Ensuring safety
- Ensuring product quality
- Improving efficiency of industrial processes
- Reduction of production costs
- Reduction of maintenance costs
- Energy saving
- Ensuring compliance with current energy and environmental standards

4.2   Temperature definition

Temperature is a physical characteristic that can be used to quantify how hot or cold a body is. There are different units for measuring temperature of a body. These include degree Celsius, degree Fahrenheit, Kelvin and absolute centigrade. It is possible to make a conversion between any pair of units mentioned. The definition of each of these temperature measuring units is different from others. For example:

i. **Kelvin (°K):** The proportion 1/276.16 of the thermodynamic temperature of the triple point of water is the Kelvin (°K) unit of measurement for temperature. The temperature at which water exists in its gaseous, liquid, and solid states in thermodynamic equilibrium is known as the "triple point of water."

ii. **The Celsius:** Water freezes at 0°C and boils at 100°C on the Celsius temperature scale (°C).

iii. **The Fahrenheit:**   The Fahrenheit unit is widely used in the United States. The freezing point of water is 32°F, and the boiling point is 211.97°F.

To convert temperature from °C to °F employ the following:

$$\text{Temperature } °F = (\textit{temperature } °C \times 1.8) + 32$$

and from °C *to* °K we use:

$$°K = °C + 273.15$$

$$Rankine\ °R = (°C + 273.15)x\ \frac{9}{5}$$

$$°C = \frac{(°F-32)}{1.8}.$$

4.3   Temperature Measuring Systems and Devices

Different systems are employed in the measurement of temperature. The following are commonly used temperature measuring systems and devices.

     i.   Thermocouples
    ii.   Resistance
   iii.   Thermistors
   iv.   Bimetallic
    v.   Pyrometers
   vi.   Quartz Crystals

4.4   Thermocouples

When two metals with dissimilar work functions are brought connected and kept at two distinct temperatures—referred to as cold and heat junctions—a thermocouple is created. At the junction, an almost proportionate electromagnetic field (emf) is produced. When two dissimilar metals are brought together at two different temperatures $(T_1 > T_2)$, the emf E that results can be expressed as

$$E = a(T_1 - T_2) + b(T_1 - T_2)^2$$

where the constants a and b have values that vary depending on the metals used.

Let $T_1 T_2 = \Delta\theta$

$$\therefore E = a(\Delta\theta) + b(\Delta\theta)^2.$$

"a" has values from 40 to 50 μv per °C and "b" few tenth or hundredth of a $\mu v/(°C)^2$.

Example

If a thermocouple instrument is made from two metals maintained at 0°C and 100°C respectively and the thermoelectric constant are 50μV/°C and 105μv/°C, calculate the emf generated at the metals junction.

Solution

$$a = 50\ \mu v/°C \qquad b = 105\ \mu v/°C$$
$$\Delta\theta = 100 - 0 = 100°C$$
$$E = a(\Delta\theta) + b(\Delta\theta)^2$$
$$= 50\ \mu v\ x\ 100 + 105\ \mu v\ x\ 100^2$$
$$= 50\ x\ 10^{-6}\ x\ 100 + 105\ x\ 10^{-6}\ x\ 10^4$$
$$= 1.055V.$$

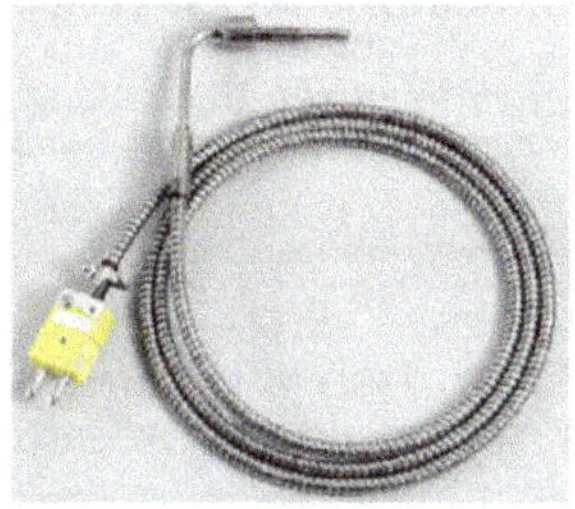

A practical Thermocouple

4.4.1   Application of Thermocouple for Temperature Measurement

A moving coil instrument with a deflection proportional to the emf is used to measure the emf generated in a thermocouple. Since temperature $\Delta\varnothing$ determines emf, the device can be calibrated to read temperature. This layout is shown in the following diagram.

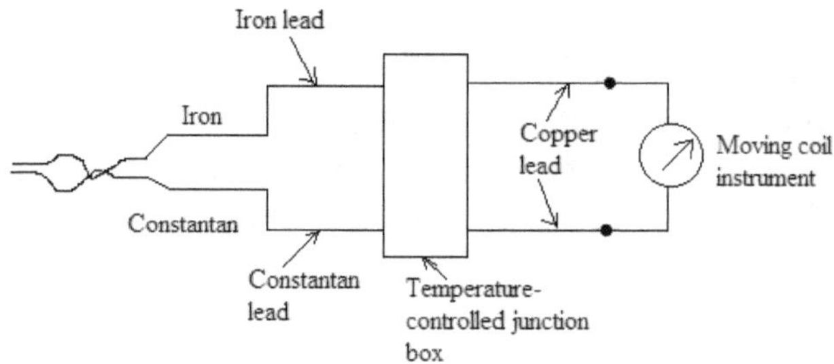

As shown here, iron and constantan metals are used. Other dissimilar metals may be used e.g., platinum and rhodium. The pair of dissimilar metals may be twisted screwed, peened, clamped or welded together. The most commonly used method is to weld the metals together.

4.4.2　Advantages of thermocouple measuring system

The thermocouple temperature measurement system has a number of advantages as listed below.

1. As opposed to resistance thermometers, thermocouples are less expensive.
2. Because thermocouples respond to temperature changes quickly, they can be used to record relative quick temperature changes.
3. Thermocouples are incredibly useful for determining the tem perature at a specific location within an equipment.

4.4.3　Disadvantages of thermocouple measuring system

The following are the disadvantages of the thermocouple temperature measuring system.

1. Because thermocouples are less accurate, they are not suitable for use in precision applications.

2. The working environment can have an impact on thermocouple metals. Precious metals such as platinum or its alloys are utilised, and they should be placed in an open-ended or closed-end metal protective tube or well to prevent contamination of the device.

3. Because the thermocouple is situated far from the measuring device, wire extension and compensating wires—which might be made of the same material as the thermocouple wires—are required, adding to the complexity of the circuitry.

Examples

A chromel-alumel thermocouple, used in thermocouple circuits, provides an emf of 33.3 mV while reading 800 °C at a reference temperature of 0 °C.

Full scale deflection is achieved with a current of 0.1 mA and a resistance of 50Ω in the meter coil R_m. The resistance between the leads and junctions R, is 12Ω. Calculate

a) Resistance of the series resistance if a temperature of 800°C is to give f.s.d.

b) The approximate error resulting from rise of 1Ω in R_c.

c) The approximate error caused by the meter's copper coil temperature rising by 10°C. The coil's resistance temperature coefficient is 0.00426/°C.

Solution

a) $\quad$ Emf $E = i_1(R_m + R_s + R_e)$

$\Rightarrow 33.3 \times 10^{-3} = 0.1 \times 10^{-3}\,(\,50 + R_s + 12)$

$R_s = 271\Omega$

b) $\quad i_2 = \dfrac{E}{R_m + R_s + R_{e1} + R_{e2}}$

$$= \frac{33.3 \times 10^{-3}}{50 + 271 + 1 + 12}$$

$$= 0.0997 mA$$

$\therefore$ Approximate error in temperature is $= \dfrac{0.0997 - 0.1}{0.1} \times 800$

$$= -2.4°C$$

c) Change in resistance of coil with temperature increase
   $of\ 10°C = 50\ x\ 0.00426\ x\ 10$

$$= 2.13\Omega$$

Current in the circuit now is $= \dfrac{33.33\ x\ 10^{-3}}{50+2.13+271+12}$

$$= 0.09936m$$

$\therefore$ Approximate error in temperature $= \dfrac{0.09936-0.1}{0.1} \times 800$

$$= -\ 5.12°C.$$

4.5   Temperature Measurement Using Quartz Crystal

The basis behind the quartz crystal thermometer is the piezoelectric property of quartz crystal, which causes an electrical charge to be generated in the material when pressure is applied to the crystals. When pressure is applied to a crystal, positive and negative charges gather like two surfaces. A capacitor with positive and negative charges on opposite sides is called a crystal oscillator. This charge is proportional to all pressures.

V=Q/C,

which is equivalent to generating a voltage.

The early crystal temperature-frequency sensors consisted of crystals with nonlinear temperature-frequency characteristics. After the discovery of crystals with linear temperature-frequency characteristics, crystals of different shapes and sizes are offered by different manufacturers with different performance indicators. These indicators include resonant frequency, resonant mode, load capacitance, series impedance, case capacitance, and drive level. It's the same one used as a frequency stabilization component. A crystal equivalent circuit is a circuit in which a capacitor is connected in parallel with the series combination of an inductor, a capacitor, and a resistor.

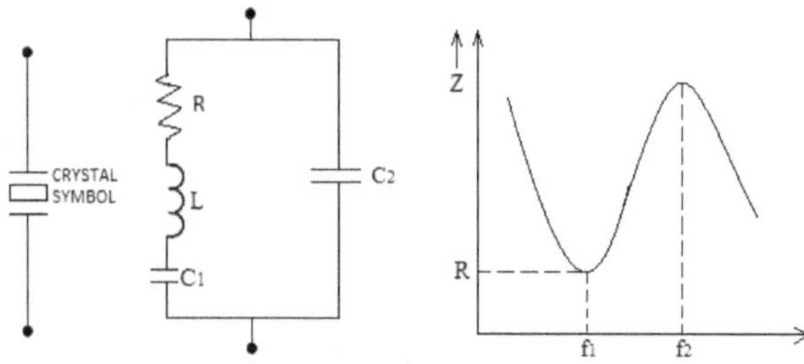

Crystal equivalent circuit and characteristics

4.6   Construction and Operating Principles of Quartz Crystal Thermometer

The **quartz thermometer** detects temperature by measuring the frequency of a quartz crystal oscillator. It is a is a high-precision, high-accuracy temperature sensor.

The temperature-sensing crystal of the thermometer is contained in a tiny sensor probe that is wired to the oscillator. Temperature readings can be taken outside of the cabinet even though the oscillator is housed in the main cabinet because it can be physically removed as a unit. Aside from that, the cabinet essentially houses a unique frequency counter that shows the recorded temperature right on. Temperature readings are taken automatically. For chemical stability, stainless steel is used to make the probe shell. The quartz wafer is positioned 0.01 inch away from and parallel to the probe's edge..

The measured temperature is effectively reduced to the observed oscillator frequency because the oscillator has a carefully cut crystal that offers a linear temperature coefficient of frequency. It is possible to obtain an accuracy of 0.02 °C and a precision of 0.0001 °C between 0 and 100 °C. High linearity provides high accuracy over the critical temperature range, which includes only one triple point of water, a convenient temperature reference point for calibration. Quartz thermometers have an accuracy of ±0.075 °C and can measure temperatures between -80 and 250 °C with precision of 0.0001°C and a sensitivity of 1000 Hz/°C.

4.6.1   Advantages of Using Quartz Crystal Thermometer

- They possess a high linear output.
- Long-term stability and reliability.
- High resolution of order 0.001°C.
- Excellent repeatability.

4.6.2   Limitations of Using Quartz Crystal Thermometer

- Low-temperature range.
- Crystals have strong cross.

4.7   Resistance Method of Temperature Measurement

The concept of resistivity underlies the operation of the resistance method of measuring temperature. Temperature variations in a material cause changes in its resistivity, which in turn causes changes in resistance. Temperature measurement may be done using this feature. This operation of the resistance thermometer is based on this principle. Recall that a metal conductor's resistance is directly proportional to its length and inversely proportional to its cross-sectional area. That is,

$$R = \frac{\rho l}{A}$$

where R = resistivity of the metal wire

l = length of the wire, m

A = Cross-sectional area of the wire, m^2

ρ = resistivity of the material of the wire, Ω-m.

An electrical resistive transducer can be created by varying any one of the values in the preceding equation. Two commonly used devices are potentiometers and strain gauges.

4.8   Construction of Resistance Thermometer

The resistance thermometer is commonly constructed of platinum metal. The choice of platinum is because it has outstanding stability and can tolerate high temperatures. It also shows limited susceptibility to contamination. The following figure shows the constructional features of an industrial platinum resistance thermometer.

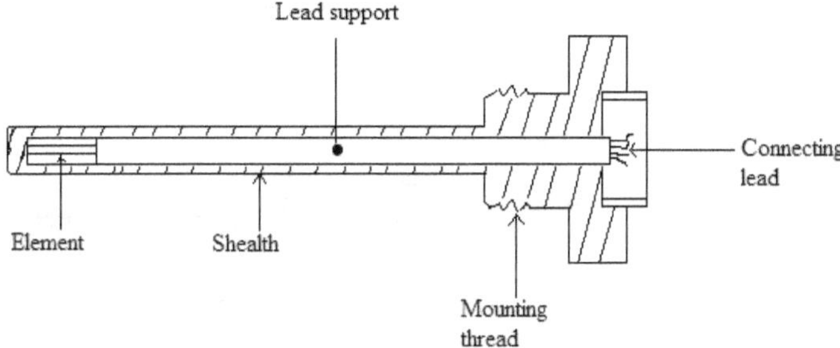

4.9   Thermistor as Temperature Measuring Device

The word "thermistor" is an abbreviation of "thermal resistor." Semi-conductor materials make up most thermistors. Since most thermistors have a negative temperature coefficient of resistance, as the temperature rises, so does their resistance. Some exhibit an extremely high temperature coefficient of resistance, which enables them to sense minuscule temperature variations. Thermistors are therefore very helpful for accurate temperature control and measurement.

4.10   Construction of Thermistors

Thermistors are made from a sintered mixture of metallic oxides, including manganese, nickel, cobalt, copper, iron, and uranium. They come in an assortment of forms and sizes. A few of these are depicted in the diagram that follows. The smallest type of thermometer is a bead, which can range in diameter from 0.0015mm to 1.25mm. To create probes that might be simpler to install than beads, beads can be sealed into the tips of solid glass rods..

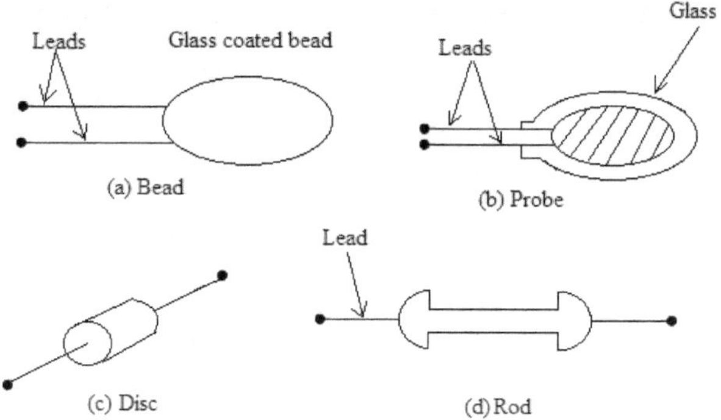

A thermistor resistance varies according to the relation:
$$R_\theta = R_\theta[1 + \alpha\theta_0\Delta\theta]$$
$$R_\theta = \text{approximate resistance at } \theta°C, \Omega$$
$$R_{\theta 0} = \text{approximate resistance at } \theta_0°C, \Omega.$$
$$\Delta\theta = \theta - \theta_0 = change\ in\ temperature, °C$$
$$\propto_{\theta_0} = \text{temperature coefficient of resistance}$$
$$\text{at } \theta°C \text{ unit is } /°C.$$

4.11   Application to temperature Measurement

Any change in temperature causes a thermistor connected in a simple series circuit with a battery and micro-ammeter to change in resistance, which in turn changes the circuit current. With a resolution of up to 0.1°C, the micro-ammeter can be directly calibrated for temperature. Usually, a bridge circuit is employed in a measuring thermistor circuit to achieve higher sensitivities. For a 4 kΩ thermistor in the following figure, a sensitivity of up to 0.005°C in temperature is achieved.

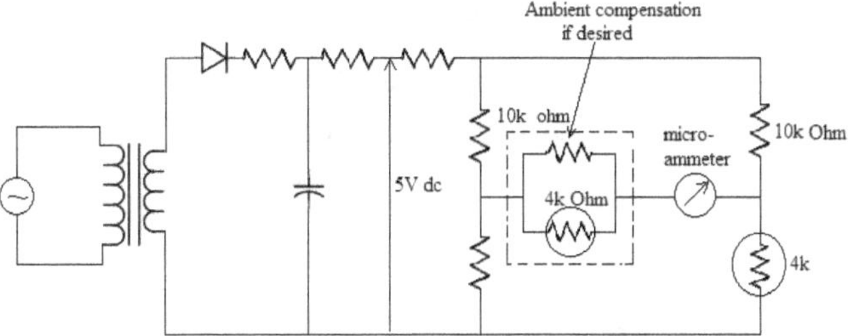

Examples
A thermistor within the temperature range of 25°C to 50°C has a resistance temperature coefficient of -5%. Determine the thermistor's resistance at 35°C if it is 100Ω at 25°C?

Solution

$$R_\theta = R_\theta[1 + \alpha\Delta\theta]$$

At35°C $\qquad R_{35} = 100\,[1 - 0.05\,(35 - 25)]$
$$= 50\Omega.$$

1) At the ice point (0°C), a thermistor's resistance is 3980 Ω, and at 50°C, it is 794 Ω. The resistance-temperature relationship is given as $R_T = aR_o\exp(\frac{b}{T})$

a) Determine the values of the constants a and b.

b) Find the range of resistance to be measured if the temperature changes from 40°C to 100°C.

Solution

a) $R_o = 3980 \, \Omega$

Absolute Temperature at ice point = 273K

$$R_T = aR_o \exp\left(\frac{b}{T}\right)$$

$$3980 = a \times 3980 \, \exp\left(\frac{b}{273}\right)$$

$$\Rightarrow 1 = a \, \exp\left(\frac{b}{273}\right) \qquad \text{(i)}$$

Resistance at 50°C is $R_T = 794$

Absolute temperature at 50°C is T

$$= 273 + 50 = 323K$$

$$\therefore 794 = a \times 3980 \, \exp\left(\frac{b}{323}\right)$$

$$\Rightarrow 0.1995 = e^{\frac{b}{273}} \qquad \text{(ii)}$$

When solved simultaneously we have:

$$a = 30 \times 10^{-6} \; and \; b = 2843.$$

b)   Absolute temperature at 40°C = 273 + 40 = 313K

$$R_{40} = 30 \times 10^{-6} \times 3980 \times \exp\left(\frac{2843}{313}\right)$$

$$= 1051.3 \, \Omega.$$

Absolute temperature at 100°C = 273 + 100

$$= 373K \; platinum$$

$$R_{100} = 30 \times 10^{-6} \times 3980 \times \exp\left(\frac{2843}{373}\right)$$

$$= 243.3 \, \Omega.$$

Range of resistance measured is 243 Ω to 1051 Ω.

2. At 0°C, the resistance of a platinum resistance thermometer is 100 Ω, and its value of α is 0.00385. The resistance when it is put to use

for temperature measurement is 105 °C. Determine the temperature measured.

Solution

In order to make θ the subject and evaluate, we reorder the formula thus:

$$\Delta\theta = \frac{\frac{R_\theta}{R_{\theta 0}} - 1}{\alpha}$$

$$\Delta\theta = \frac{\frac{105}{100} - 1}{0.00385}$$
$$= 12.987 \ °C.$$

4.12   Bimetallic Thermometers

These are widely used for measuring local temperatures in the process industries. Two basic principles are utilised by bimetallic thermometers:
i) All metals undergo thermal expansion or contraction.
ii) Metals expand and contract at varying rates because not all metals have the same temperature coefficient of expansion. The deflections corresponding to temperature change are produced by using the difference in thermal expansion rates.

4.13   Construction of Bimetallic Thermometer

The component of a bimetallic thermometer is a bimetallic strip, which is made by joining two thin strips of two distinct metals so that they are immobile with respect to one another. Following is an illustration of this:

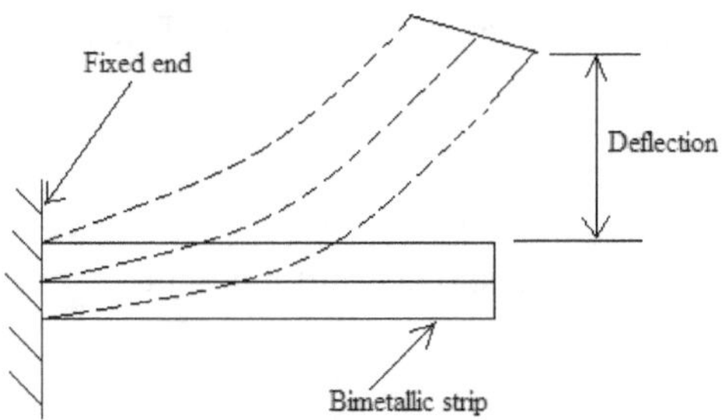

The two metallic strips change length at different rates because, when exposed to the same change in temperature, all metals attempt to change their physical dimensions at different rates. The bimetallic strip bends in response to temperature changes because of the two metals' differing rates of expansion. Throughout the linear portion of the deflection-temperature characteristics, the bimetallic strip's free end's deflection is inversely proportional to thickness and proportional to temperature changes. With the right transducer, such as a variable resistance, inductance, or capacitance, the deflection of the free end can be transformed into an electrical signal. Extending the length of the strip can boost a bimetallic thermometer's sensitivity. The deflection of the free end of the bimetallic strip is given as

$$r = \frac{t[3(1+m)^2+(1+mn)\left(m^2+\frac{1}{mn}\right)]}{6\,(\alpha_A-\alpha_B)(T_2-T_1)(1+m)^2}$$

where r = radius of the arc of deflection

t = total thickness of strip

n = ratio of moduli of elasticity $= \dfrac{E_B}{E_A}$

m = ratio of thickness $= \dfrac{t_B}{t_A}$

$T_2 - T_1$ = change in temperature

t_A, t_B = thickness of metal A and B respectively

$\alpha_A\ \alpha_B$ =thermal coefficients of expansion of metals A and B respectively.

In most practical applications, the metals and their dimensions are chosen so that moduli of elasticity and thickness are equal i.e.,

$$E_{B=}E_A \text{ or } n=1$$
$$\text{and } t_{A=}t_B \text{ or } m=1.$$

In that case r has the expression

$$r = \frac{2t}{3(\alpha_A - \alpha_B)(T_2 - T_1)}.$$

In case one of the metals has a very low thermal expansion coefficient, say metal B, we have $\alpha_{B \approx 0}$

$$\therefore r = \frac{t}{2\alpha_A(T_2 - T_1)}$$

Examples

1. Strips of nickel-chromium alloy and invar bound together at 25°C make up a bimetallic thermometer. Each metal strip is 50 mm long and has a thickness of 1 mm. Determine the radius of curvature that results from exposing the strip to a temperature of 200°C without any constraints. The coefficients of expansion and moduli of elasticity for nickel chrome alloy and Invar are: $216GN/M^2$, $147GN/m^2$ and $12.5\times 10^{-6}/°C$, $1.7\times 10^{-6}/°C$.

Solution

$$m = \frac{t_A}{t_B} = 1, t = 2 \times 1 = 2mm$$

$$n = \frac{E_B}{E_A} = \frac{147}{216} = 0.68$$

Change in temperature, $T_2 - T_1 = 200 - 25 = 175 \ °C$.

$$r = \frac{t[3(1+m)^2 + (1+mn)\left(m^2 + \frac{1}{mn}\right)]}{6(\alpha_A - \alpha_B)(T_2 - T_1)(1+m)^2}$$

$$r = \frac{2[3(1+1)^2 + (1+0.68)\left(1^2 + \frac{1}{0.68}\right)]}{6(12.5 \ x10^{-6} - 1.7 \ x10^{-6})(175)(1+1)^2}$$

$$= \frac{1}{3}\left(\frac{16.1506}{0.00756}\right)$$

$$= 711 \text{ mm.}$$

2. A bimetallic strip element features a 40 mm cantilever with one fixed end and the other free. Each metal has a thickness of 1 mm,

and at 20 °C, the element is initially straight. Determine how much the free end will move from the starting line in a perpendicular direction if the temperature is raised to 180 °C. A nickel-chrome alloy has an expansion coefficient of 12.5 $X10^{-6}/°C.$, while Invar, one of the metals, has a negligible thermal expansion coefficient.

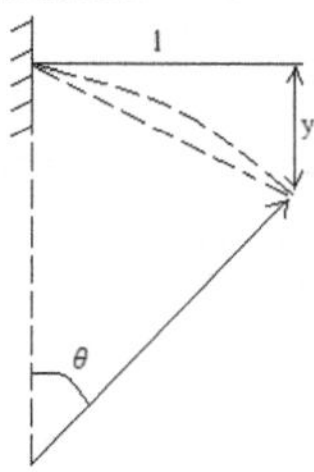

Solution

t = 2 x 1 = 2mm

Since $\propto$ for one metal is negligible

$$r = \frac{t}{2(T_2 - T_1)\propto_A}$$

$$= \frac{2}{2(180-20)\ x\ 12.5\ x\ 10^{-6}}$$

= 500mm.

From the figure,

$$\theta = \frac{l}{r} = \frac{40}{500} = 0.08 \text{ rad}$$

$$\theta = 4.58°$$

$$\therefore y = r(1 - Cos\theta) = 1.6\text{mm}$$

4.14   Measurement of Temperature by Pyrometry

Thermal radiation or optical pyrometers are used when temperatures are high and it is impractical or impossible to make direct contact with the process to be measured. When corrosive liquids or vapours come into contact with the medium being measured, they can ruin thermocouples, resistance thermometers, and thermistors. In these situations, pyrometers are utilised. These pyrometers are useful for measuring temperatures for items that move quickly and for temperatures above the thermocouple's detection range.

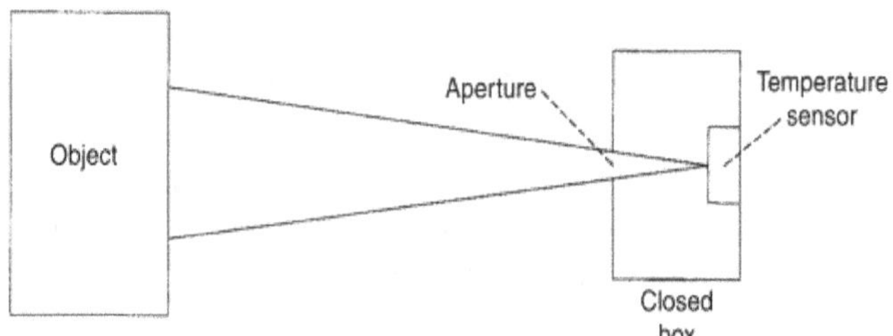

The radiant heat that a heated object emits or reflects is measured using radiation pyrometry. Electromagnetic radiations that fall between 0.1 and 100μm in wavelength are known as thermal radiations.

Pyrometers work on the basis of black body theories. A black body's total thermal radiation output is expressed as:

$$q_b = \sigma T^4 \, W/m^2$$

$$\sigma = stefan\ Boltzmann\ constant = 572 \times 10^{-9} w/m^2$$

$$T = absolute\ temperature,\ K.$$

In comparison to a rough black surface, a smooth brilliant surface radiates less heat. The term "emissivity" ε for this phenomenon is as follows:

$$\varepsilon = q/q_b$$

$$q = heat\ radiated\ by\ a\ gray\ body,\ W/m^2$$

$$q = \varepsilon \sigma T^4 w/m^2$$

from which

$$T = (q/\varepsilon\sigma)^{¼}$$

Example 1

An optical pyrometer's clouded lens results in a transmission factor of 0.8. The temperature on the instrument is 1480°C. Determine the actual temperature?

Solution

Apparent absolute temperature, $T_a = 1480 + 273 = 1753K$

True absolute temperature, $T = \varepsilon^{-\frac{1}{4}}T_a$
$$= (0.8)^{-\frac{1}{4}} \times 1753$$
$$= 1853.7K$$
True temperature $= 1853.7 - 273$
$$= 1580.7°C.$$

2) When the temperature of a metal object is tested, it is found to be 1065°C, assuming a surface emissivity of 0.82. Subsequent research revealed that the actual emissivity is 0.75. Determine the measurement error for temperature.

Solution

With $\varepsilon = 0.82$, absolute $T = 1065 + 273 = 1338K$
Apparent temperature $T_a = (0.82)^{\frac{1}{4}} \times 1338$
$$= 1273.2 \text{ K}$$
Actual absolute temperature with $\varepsilon = 0.75$
is, $T = \varepsilon^{-\frac{1}{4}}T_a$
$$= (0.75)^{-\frac{1}{4}} \times 1273.2$$
$$= 1368.2K$$
Actual temperature $= 1368.2 - 273$
$$= 1095.2°C$$
$\therefore$ Error in temperature measured $= 1095 - 1065$
$$= 30°C.$$

4.15   Instrument Calibration

Calibration of measuring devices is required to guarantee measurement reliability. The process of calibrating a measurement device involves recording its comparison with a traceable reference standard or device. A "calibrator" is another term for the reference standard. The reference standard should logically be more accurate than the instrument to be calibrated in order to produce the desired results. It is also necessary to calibrate the reference standard traceably. In applications involving manufacturing processes, measuring apparatus needs to be calibrated several times over its operational range in order to provide accurate data for crucial alarms and systems. A poor calibration or a failure to calibrate can cause harm, financial losses, fatalities, and even significant environmental catastrophes.

4.15.1 Purpose of calibration

There are three main reasons for having instruments calibrated:

1. To ensure measuring instrument readings are consistent with other measurements.

2. To determine the accuracy of the instrument readings.

3. To establish the reliability of the instrument.

The term calibration refers to the **accuracy** and **quality** of the measurements that are made on the apparatus. Measurement equipment's accuracy and results often "drift" with time. It is always necessary to maintain instrument calibration for measurement results. Measurements made with calibration are repeata**ble, accurate, and dependable.** Calibration is used to ensure measurement accuracy with the goal of minimizing measurement uncertainty. Errors or uncertainties in measurements are quantified and controlled by calibration.

4.15.2 Thermometer Calibration

A thermometer should be calibrated when newly designed (before use), when dropped, when changing from one temperature range to another and after a longer storage period. For most applications, the thermometer should be within $\pm$ 0.5°C of the reference thermometer used for calibration. To calibrate a thermometer, its accuracy must be tested first with a substance of known temperature. Then the thermometer temperature is set to that temperature.

One method of calibrating thermometers is the use of calibration baths. A calibration bath is an electronic controller that provides temperature automatically and quickly with the help of liquids.

Liquid is used in a calibration bath to automatically measure temperature values that are within the probe's temperature range. There are two types of calibration baths: big laboratory rigs that are permanently fixed, or portable devices. Depending on the temperature to be monitored, the probe is submerged in either a bath of water or a certain kind of oil. The calibration bath's water or oil circulates to maintain the target temperature. Following measurement, the outcomes are contrasted with a reference tool. The utilisation of calibration baths enhances temperature consistency.

The calibration bath's great dependability, accuracy, and remarkable consistency within the measurement chamber make it an excellent choice for use as a factory or operational standard for automated testing and/or calibration of different temperature sensors, regardless of diameter.

A portable temperature calibration bath

Exercise

1. a) Define temperature.
b) State and explain four measurement units of temperature.
c) Convert the following temperature:
      i)   85°F to Kelvin
      ii)   650°F to Kelvin
      i)   25K to degree Celsius
      ii)   75K to degrees Fahrenheit
      iii) 1098°C to °F
      iv) 135 K to °F
      v)   249°Fto °R
      vi) 450°R to °C.

2. a) Explain the principle of operation of the thermocouple.
b) With the aid of diagram, explain the application of thermocouple for temperature measurement.
c) A thermocouple instrument is made from two metals maintained at 0°C and 100°C respectively and the thermoelectric contains are

$55\mu V/°C$ and $125\mu V/°C$. Determine the emf produced at the junction of the two metals.

d) State and explain:

(i) three advantages and (ii) three disadvantages of using the thermocouple to measure temperature.

3. a) State and explain one resistive method of temperature measurement.

b) Distinguish between the principle of operation of a thermocouple and that of a thermistor

c) Explain with the aid of diagram, the construction of a thermistor.

d) How would you use a thermistor to measure temperature?

4. Explain the construction and the principle of operation of the following temperature measuring devices:
   a. Thermocouple
   b. Thermistor
   c. Resistance thermometer
   d. Quartz crystal thermometer
   e. Bimetallic thermometer

5. Explain how two metals bonded together can be employed to measure local temperature.

b) A strip made of two metals is used as a thermometer. If the metals are NiCr alloy and Invar each having a thickness of 0.9mm are at 23°C, calculate the curvature produced with a temperature of 240°C. The modules of elasticity and coefficient of expansion are given respectively as $216GN/m^2$, $147GN/m^2$ and $12.5 \times 10^{-6}/°C$.

6. Show that the radius of curvature of a restrained bimetallic strip having one metal having very low thermal expansion coefficient and the metals having equal thickness and modulus of elasticity can be given as

$$r = \frac{t}{2\alpha_A(T_2 - T_1)}.$$

(ii) Hence calculate the perpendicular displacement of a bimetallic strip with one free end if the strip was initially free at 20°C and the temperature is later raised to 240°C. The thickness of each metal is 1.1 mm and the length of cantilever is 50 mm. One metal has negligible thermal coefficient of expansion while the other metal is NiCr with expansion coefficient of $12.5 \times 10^{-6}/°C$.

7. Explain the process of:
   a) instrument calibration.
   b) thermometer calibration.

8. When the temperature difference between the sensor tip $\theta1$ and the gauge head $\theta2$ is equal to e $= \alpha(\theta_1-\theta_2) + \beta(\theta_1^2-\theta_2^2)$ and $\alpha = 3.5 \times 10^{-2}$ and $\beta = 8.2 \times 10^{-6}$, a thermocouple generates a 12 mV e.m.f. The temperature is 20°C at the gauge head. Determine the temperature registered at the sensor. Ans. 20.33°C.

9. If a thermocouple instrument made from two metals gives the voltage of 1.35V and 1.055V when maintained at (0°C and 100°C) and (20°C and 105°C) respectively, determine the thermoelectric constants.

CHAPTER FIVE

PRESSURE MEASUREMENT

5.1   Introduction

The force per unit area delivered in a direction perpendicular to an object's surface is known as pressure. It may also be described as the force exerted on a surface's unit area. From a mathematical standpoint,

$$P = F/A$$

We refer to this as absolute pressure.

5.2   Pressure Units and Conversion

Pressure is measured in Pascals, *Pa* which is equal to one Newton per square meter. That is

$$1\ Pa = 1\ N/m^2$$

Other units are bar, torr and atmosphere (atm). The units can be converted one from the other using the following relations:

$$1Pa = 1N/m^2 = 10^{-5}bar = 9.8692 \times 10^{-6}atm = 7.5006 \times 10^{-3}torr$$

1 torr is the same as millimeter of mercury.

5.3   Forms of Pressure

The force per unit area that the weight of the air above a surface in the earth's atmosphere exerts against it is known as **atmospheric**

pressure. The hydrostatic pressure generated by the weight of air above the measuring location is a good approximation of atmospheric pressure.

When considering the load on storage containers and the piping elements of fluidics systems, **gauge pressure** is the pertinent measurement of pressure.

Gas pressure is measured in relation to **atmospheric pressure** rather than in absolute terms. Gauge pressure refers to these types of pressure measurements. For instance, the approximate 220 kPa of air pressure in a vehicle tire is actually 220 kPa higher than atmospheric pressure. The absolute pressure in the trench is consequently around 320 kPa, as the air pressure at sea level is approximately 100 kPa. As a result, gauge pressure is the measurement of pressure in relation to ambient pressure or the local atmosphere.

Differential pressure refers to the measured difference between two distinct, unrelated bodies. Additionally, it can refer to the differential in pressure that occurs between two points in a system, as the inlet and output of a pump. A system is said to be in vacuum pressure if its pressure is lower than that of the atmosphere.

$$P = P_g + P_o$$

Where P = absolute pressure

P_g = gauge pressure,

P_o = atmospheric pressure

In terms of measurement, **absolute pressure** is the pressure in relation to a vacuum. As previously mentioned, the measurement of pressure is typically done in relation to the atmospheric pressure.

Gauge pressure below atmospheric pressure is indicated as **vacuum pressure**. This is calculated as a pressure value that is negative. If the pressure is positive, it is measured above atmospheric pressure.

Examples

1) When the local atmospheric pressure is 98kPa (Abs), determine a pressure of 155kPa (gauge) as an absolute pressure.

Solution

$P_a = P_g + P_o$
$=155KPa + 98KPa$
$=253KPa$ (abs).

2) Given that the local atmospheric pressure is 103.4KPa, convert a pressure of 75.2KPa to a gauge pressure.

Solution

$P = P_g + P_o$

$P_g = P - P_o$
$=75.2-103.4=-28.2kPa$ or $28.2kPa$ vacuum

5.4   Pressure measurement using manometers

A static fluid's change in elevation and pressure are related, and this relationship is used by the manometer as a pressure measurement device. This is expressed as

$$\Delta P = \gamma h$$
$$\text{where} \quad \Delta P = change\ in\ pressure$$
$$h = \text{change in elevation.}$$

The following figure shows a sketch of typical manometer. The device is in form of a U-tube filled with a gauge fluid, water and an air column as shown.

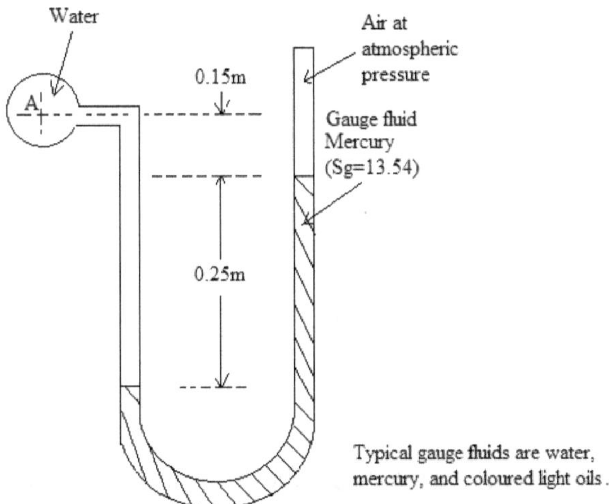

The gauge fluid is forced out of its partial location by the pressure that has to be measured. Since the fluids inside the manometer are at rest, formulas for the pressure changes that happen throughout the device may be written using the equation $\Delta p = \Delta h$, which can then be solved algebraically to get the desired pressure.

Example 1: Calculate the pressure at point A in the figure below.

Solution

Pressure at point 1 is known. This is P_1.

The pressure that exists within the mercury, 0.25m below points 1 is given as

$$P_1 + \gamma_w(0.25m)$$

where $\gamma_m(0.25m)$ = the change in pressure between point 1 and 2   due to a change in elevation.

γ = specific weight of mercury, the gauge fluid.

We now write an expression for point 3 in the left leg as:

$$P_1 + \gamma_w(0.25m) - \gamma_w(0.4m)$$

where $\gamma_w(0.4m)$ = the specific weight of water.

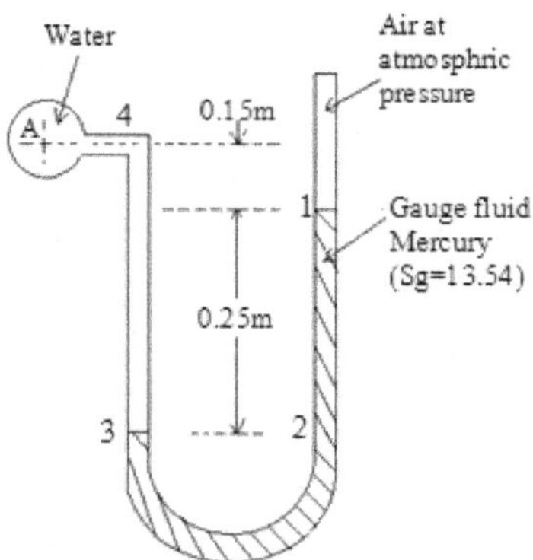

The last term is subtracted from the previous expression because between points 3 and 4, the pressure decreases. The pressure at the point A is equal to that of point 4 since both points are at the same level.

$$P_A = P_1 + \gamma_{m(0.25m)} - \gamma_{w(0.40m)}$$

Now $P_1 = P_{atm} = 0Pa$ (gauge)

$$\gamma_m = (3g) \text{ m x } 9.81kN/m^3$$
$$= (13.54)(9.81)$$
$$= 132.8kN/m^3$$
$$\gamma_w = 9.81kN/m^3$$

$$P_A = P_1 + \gamma_{m(0.25m)} - \gamma_{w(0.40m)}$$
$$= 0Pa \text{ (gauge)} + 132.8KN/m^3 \text{x}$$
$$0.25m - 9.81KN/m^3 (0.40m)$$
$$= 0Pa + 33.20kN/m^2 - 3.92kN/m$$
$$P_A = 29.28kN/m^2$$
$$= 29.28kPa \text{ (gauge)}.$$

2. Determine the pressure differential between points A and B in the given figure and represent it as $P_A - P_B$. The differential manometer gets its name from the fact that it measures the difference in pressure between two points, not the actual value of either one.

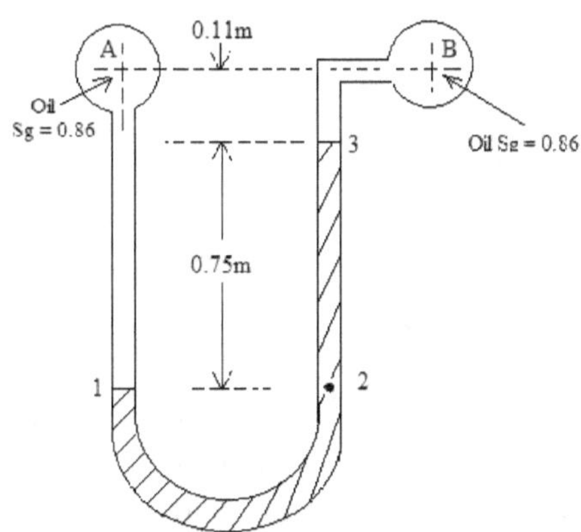

Solution

Let P_A = Pressure point A

P_B = Pressure point B

$$P_1 = P_A + \gamma_{o(0.86m)}$$

where γ_o = specific weight of the oil

Pressure at point 2, P_2 is determined as

$$P_2 = P_1 + \gamma_{w(0.75m)}$$
$$= P_A + \gamma_{o(0.86m)} + \gamma_{w(0.75m)}$$

Now pressure at point 4 P_4

$$P_4 = P_A + \gamma_{o(0.86m)} - \gamma_{w(0.75m)} - \gamma_{o(0.11m)}$$

This is also the pressure at point B since points 4 and B are on the same level.

i.e. $P_B = P_A + \gamma_{o(0.86m)} - \gamma_{w(0.75m)} - \gamma_{o(0.11m)}$

$$P_B - P_A = \gamma_{o(0.86m)} - \gamma_{w(0.75m)} - \gamma_{o(0.11m)}$$

Now,

$$\gamma_o = (s_g)_o(9.81\text{kN/m}^3) = 0.86 \times 9.81 = 8.44\text{kN/m}^3$$
$$\gamma_w = 9.81\text{kN/m}^3$$
$$P_B - P_A = \gamma_{o(0.86-0.11)} - \gamma_{w(0.75m)} = \gamma_{o(0.75m)} - \gamma_{w(0.75m)}$$

$$= 0.75(\gamma_o - \gamma_w)$$
$$= 0.75(8.44 - 9.81)\text{kN/m}^3$$
$$= 0.75(-1.37)$$

$$= -1.03 \text{kN/m}^2$$

The negative sign indicates that the magnitude of P_A is greater than that of P_B.

5.5 Pressure Measurement with Bellows

As shown in the figure, a metallic bellows is made out of a sequence of circular pathways that resemble an accordion's folds. It is composed of a very ductile and strong alloy. Bellows are metallic cylinders with thin walls and deep convolutions. One end of the cylinder is sealed while the other is left open. While the open end is fixed, the closed end is free to move. Bellows are made of a variety of metals, including stainless steel, Monel, Inconel, brass, bronze, and beryllium copper. The metals need to be highly resistant to fatigue failure, thin enough to be flexible, and ductile enough for comparatively simple production.

Because of the way the parts are attached, variations in pressure might cause them to expand or shrink axially. The bellows will compress when pressure is applied to the closed-end. The link, which is the rod that runs between the bellows' closed end and the transmission mechanism, will rise together with the closed end, moving the pointer as it does so.

The low- to medium-pressure range can be measured using bellows for both absolute and differential pressure. The force that can be transferred to the transmission mechanism depends on the bellows' diameter. In order to have enough surface area on which the measured pressure can act, a bigger diameter is selected for the measurement of very low pressures.

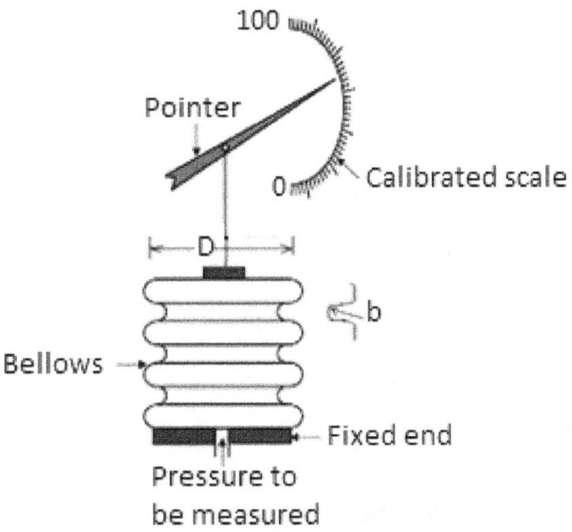

The displacement of the bellows element is given by

$$d = \frac{0.453 PbnD^2 \sqrt{(1-v^2)}}{Et^3}$$

where P = pressure, N/m^2

b = radius of each corrugation

n= number of semi-circular corrugations

t = thickness of wall, m

D = mean diameter, m

E = modulus of elasticity, N/m^2

V = Poisson's ratio.

Sometimes the movement of a bellows is countered by a calibrated spring, allowing only a portion of the maximal stroke to be used, in order to prolong its lifespan and improve accuracy. This is referred to as a spring-loaded bellows. The deflection of the spring-loaded bellows is given as:

$$d = p\ \frac{A_b}{K_b + K_s}$$

A_b = effective area of bellows, m^2

K_{b},K_{s} = stiffness constants of bellows and spring respectively, N/m^2

$$\therefore \text{Pressure } P = \frac{d(K_p + K_s)}{A_b}$$

If the bellows assembly is powered by a mechanism such as electric switches, then,

$$P = \frac{F + d_s(K_b + K_s)}{A_b}$$

where F = force required to operate switch or mechanism, N

d_s = deflection required to operate switch or mechanism, m.

5.5.1 Advantages of using bellows for pressure measurement

1. It is rugged in construction.
2. Its cost is moderate.
3. Bellows has capability of providing large force and wide pressure range.

5.5.2 Draw-backs of bellows

1. Ambient temperature compensation is required.
2. Bellows are susceptible to vibrations, hysteresis, drift, work hardening, friction, and temperature variations.
3. It is unsuitable for measuring very high pressure due to their larger bulk and more extended relative motion.

Example

A differential bellows arrangement uses two bellows each of natural length 50 mm, effective area of 1500 mm^2, and stiffness 0.5N/mm. bellows A is evacuated and consists of stiffness 3N/mm. Find the natural length of the spring if the bellows are to be equally compressed to a length 40mm when a pressure of $100KN/m^2$ absolute is applied to B. Also find the displacement of the output point C for a change of $10KN/m^2$ in applied pressure.

Solution

When a pressure of $100KN/m^2$ is applied

$$F = PA = 100 \times 10^3 \; 1500 \times 10^{-6}$$
$$= 150N$$

$\therefore$ spring compression $= {}^{150}/_3 = 50mm$

$\therefore$ Natural length of spring $= 50 + 40 = 90mm$

When a pressure of $10KN/m^2$ is applied;

$$F = 10 \times 10^3 \times 1500 \times 10^{-6}$$
$$= 15N.$$

Total stiffness of spring plus stiffness of bellows is

$$= 3 + 2 \times 0.5$$
$$= 4N/mm.$$

$\therefore$ Displacement of point $C = \dfrac{15}{4} = 3.75mm$.

5.6   Measuring Pressure with Diaphragms

Bellows and diaphragm elements both function on a similar principle. The diaphragm deflects in response to the applied pressure when the pressure to be measured is applied to it. The diaphragm's dimensions and thickness determine how it moves. Because of its tiny movement, a diaphragm element doesn't need springs, unlike bellows.

One side of the diaphragm is subjected to the unknown pressure. Because of the exerted pressure, the diaphragm's edge, which is rigidly fixed, deflects.

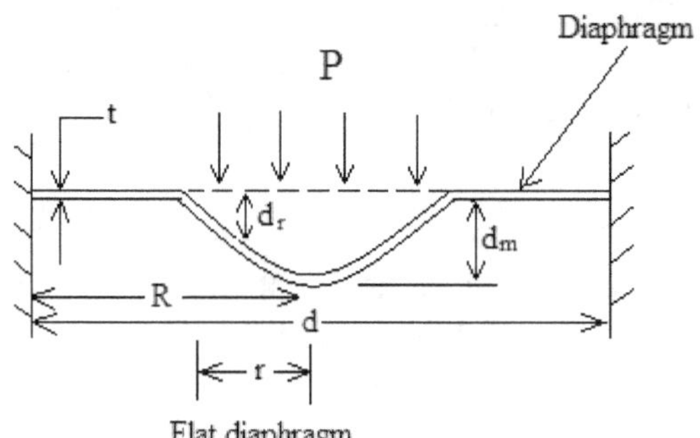

The diaphragm, a very thin membrane under pressure and hence under radial tension, can use capacitive or inductive transducers to produce an electrical output that is proportionate to the output of the transducers. The displacement of the diaphragm can be measured to determine the value of applied pressure, P. Diaphragms with thin membranes are only useful for measuring very small pressure variations. Membrane diaphragms with thin membranes are appropriate for resistive and piezoelectric transducers that need larger displacements than those produced by membrane diaphragms since membranes have a limited force-producing capacity.

5.6.1   Advantages of Diaphragm Pressure Gauges

i.   They possess excellent performance under load.
ii.   Appropriate for monitoring differential and absolute pressure.
iii.   Compact, and reasonably priced.
iv.   Suitable for measuring viscosity in slurries.

5.6.3   Disadvantages of Diaphragm Pressure Gauges

i.   Not applicable for high pressure measurement.
ii.   They possess poor seismic, impact resistance.

iii.   Difficult to maintain.

5.7   Using Bourdon Tubes to Measure Pressure

A bourdon tube is made up of a tube that has been bent into a spiral or helical shape, or flattened to have an approximately elliptical cross-section. The entire length of the bourdon is made of seamless metal tubing, with wall thickness ranging from 0.25mm to 1.25mm. The tube is sealed at one end and the other end left open for the application of pressure under measurement. The tube is soldered or welded to a socket at the base, which allows for the establishment of a pressure connection. The materials used to make bourdon tubes include phosphor-bronze, alloy-steel, cold worked brass, beryllium copper, stainless steel, and morel metal. The material employed must have good elastic or spring characteristics for adequate reliability.

The tube's open end, which is mechanically secured, allows the fluid whose pressure is being measured to enter the inside. The tube then experiences a deflection that is proportionate to the pressure's intensity. To provide an electrical signal, the deflection is mechanically conveyed to either the LVDT core or the wiper of a potentiometer, as seen in the accompanying figure.

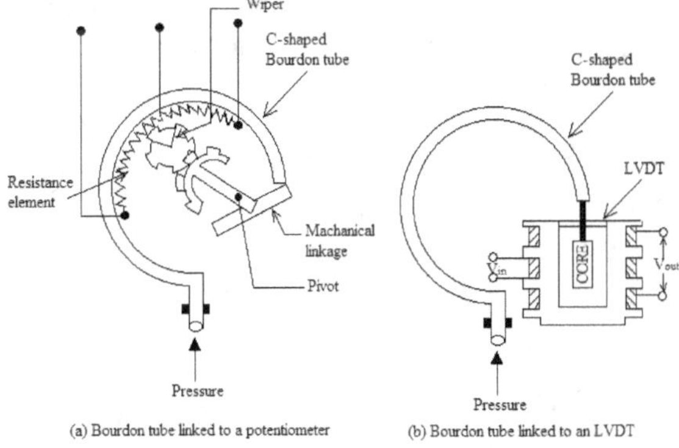

(a) Bourdon tube linked to a potentiometer   (b) Bourdon tube linked to an LVDT

5.7.1   Advantages of Bourdon Tubes

1. They have cheap cost and simple construction
2. They are available over a wide range of pressure
3. High sensitivity and good repeatability
4. They also hand good accuracy except at low pressures.
5. Bourdon tubes are easily adaptable for design for obtaining electrical outputs.

5.7.2   Disadvantages of Bourdon Tubes

1. Bourdon tubes are prone to shock vibration and shock resistance.
2. Low spring gradient limit their use for precision measurement up to a pressure of 3MPa.

5.8   Electrical Pressure Transducers

A sensor that transforms pressure into an analog electrical signal is called a pressure transducer, also known as a pressure transmitter. This conversion process involves first converting a mechanical motion into a change in electrical resistance, which is subsequently transformed into a change in electrical current or voltage. The strain-gauge kind of pressure transducer is one of the most widely used varieties. Others are:

- Piezoelectric pressure transducer
- Capacitance pressure transducer
- Inductive pressure transducer
- Potentiometric pressure transducer
- Resonant wire pressure transducer

Pressure transducers usually use a piezoresistive element, which alters its resistance proportionate to the strain (pressure) encountered.

Ventilated cables are typically used by pressure transducers in order to mitigate the impact of variations in atmospheric pressure. Pressure transducers are particularly helpful for long-term water level monitoring because they also have an automatic digital data logger. Furthermore, depending on the needs, data-recording periods can be configured anywhere from a few seconds to several days.

5.8.1 Strain gauge pressure transducer

The strain gauge pressure transducer is widely employed in numerous control systems to measure flow, level, and pump pressure. As seen below, the strain gauge in the strain gauge pressure transducer is attached to the diaphragm of the pressure transducer and connected to a Wheatstone bridge structure. With the application of pressure, the diaphragm deflects which will inevitably cause strain to be applied to the strain gauge. As a result, a change in resistance proportionate to the applied pressure occurs in the strain gauge.

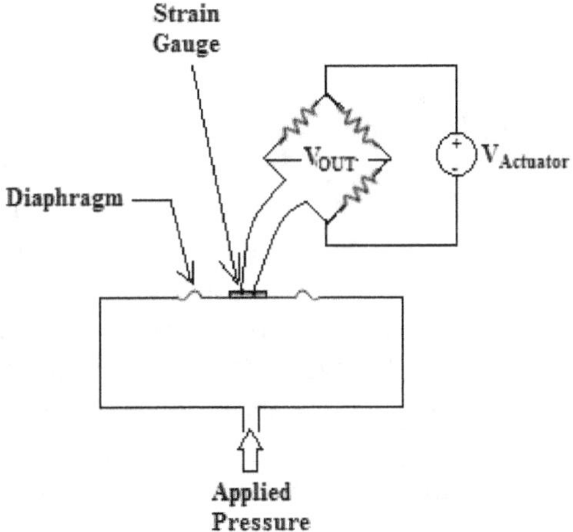

Strain gauge pressure transducer.

5.8.2   Piezoelectric Pressure Transducer

Any type of physical quantity can be transformed into an electrical signal that can be utilised for measurement using a piezoelectric transducer, which is an electrical transducer. A piezoelectric transducer is an electrical transducer that transforms physical quantities into electrical signals by utilising the characteristics of piezoelectric materials. The application of any kind of mechanical stress or strain results in a voltage that is proportionate to the applied voltage in materials with piezoelectric characteristics. A voltmeter can be used to measure the generated voltage and compute the amount of stress or strain on the material.

The materials used for measurements are accessible in a variety of shapes or without compromising their qualities, have high power values and frequency stability, and are unaffected by high or low humidity. It needs to be sufficiently flexible to be produced in a variety of shapes. For this, quartz crystals coated in silver are employed.

Principle of Operation

The basis for the operation of piezoelectric transducers is the principle of piezoelectricity. In this principle, a tiny layer of a conductive substance, like silver, is applied to the surface of the piezoelectric material, which is often quartz. Ions in the material migrate toward one conductive surface and away from the other under stress. As a result, charges are produced. The applied stress is measured using the charges that are produced. As seen below, the direction of the applied stress determines the polarity of the charge that is created.

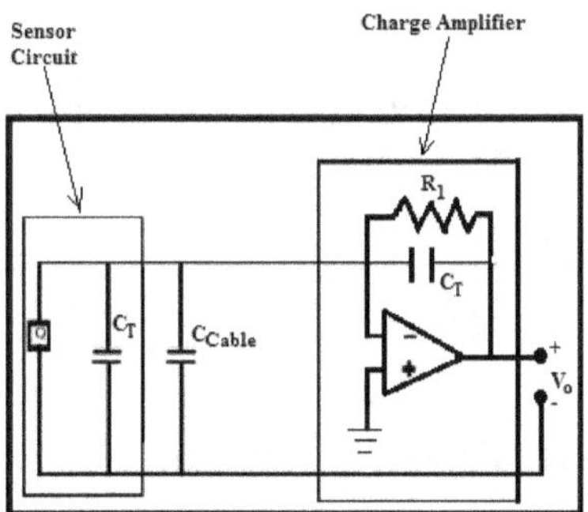

Piezoelectric Pressure Transducer

A charge amplifier is used to enhance the proportional voltage generated by mechanical stress in the piezoelectric transducer, which is then utilised to calibrate the applied voltage. When stressed, the crystal with the silver coating functions as a sensor and produces a voltage. To measure the created charge without loss, a charge amplifier is employed. R1 must be a very large resistor in order to draw very little current. Another factor influencing calibration is the capacitance of the connection connecting the piezoelectric sensor and transducer. Because of this, the charge amplifier is typically positioned somewhat near to the sensor..

Advantages of the piezoelectric pressure transducer

The following are the advantages of the piezoelectric pressure transducer:

- These are active converters.
- It is self-generated as it does not require external power to operate.
- These converters operate at high frequencies, making them suitable for a variety of applications.

Limitations

• Temperature and environmental conditions can affect converter operation.

• It is not useful for measuring static parameters as it can only measure changing pressure.

5.8.3   Capacitance Pressure Transducer

In this technique, a diaphragm (Plate 1) and an electrode attached to an unpressurized surface (Plate 2) combine to form two plates of a parallel-plate capacitor. The diaphragm and the electrode, or the two plates, are kept apart by a specific amount of space. Pressure changes the capacitance by expanding or contracting the space between the two plates. Thus, by employing an appropriate method, such as a sensitive linear comparator circuit, the change in capacitance is translated into an electric signal. This transducer can measure gauge, differential, or absolute pressure, depending on the application.

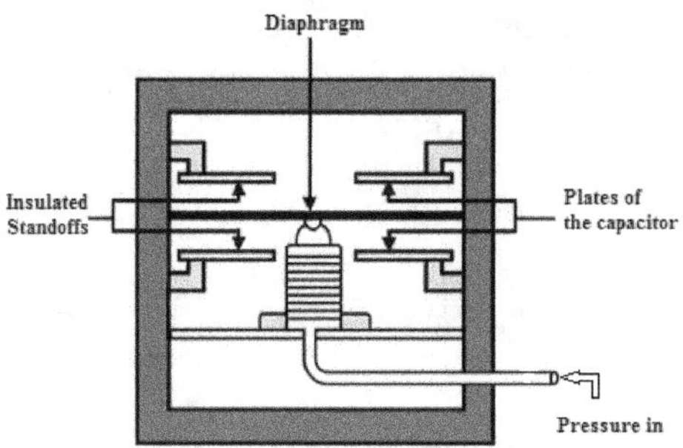

Capacitance pressure transducer

5.8.4 Inductive pressure transducer

Based on the principle of electromagnetic induction, inductive pressure transducers function by means of a membrane connected to a ferromagnetic core, as depicted in the figure. A small angle of the membrane induces linear motion in the ferromagnetic core, resulting in the induction of a current, which varies in response to pressure variations due to the core's movement. This variation in current is then transformed into a signal that can be used.

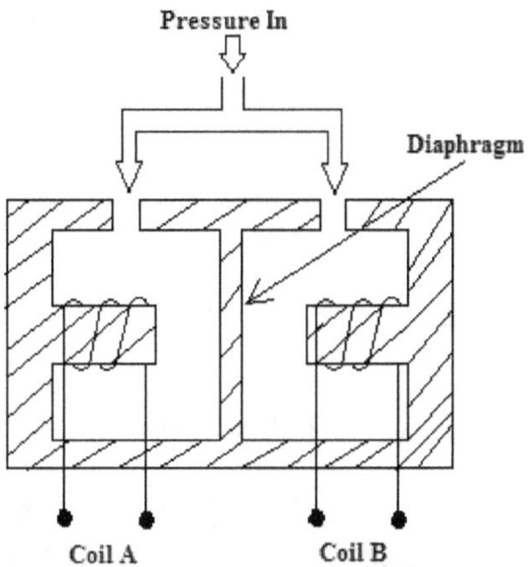

Inductive pressure transducer

5.8.5 Potentiometric pressure transducer

A straightforward technique for converting a mechanical pressure input into an electrical output is offered by the potentiometric pressure transducer. A highly precise potentiometer is the component of this kind of pressure transducer. The potentiometer is made up of a wiper that is attached to a membrane or other pressure-sensitive component. This element's deflection changes the wiper's position, which modifies the resistance between the wiper and one end of the potentiometer. The applied pressure is shown by the resistance value.

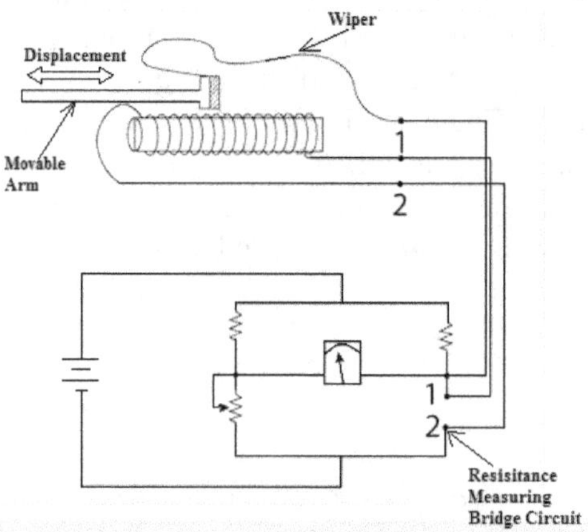

Potentiometric pressure transducer

5.8.6   Resonant wire pressure transducer

The development of resonant line pressure transducers began in the late 1970s. This instrument is frequently used to monitor pressure in industrial settings. A pressure-sensitive membrane (under tension) is attached to one end of a wire that is fixed by a static element of a resonant wire pressure transducer. Process pressure is recorded by high and low pressure membranes located on the left and right sides of the apparatus. A magnetic field is applied to the wire, causing it to vibrate. A wire oscillation at a resonant frequency is produced by the oscillator circuit.

Process pressure variations have an impact on the wire's tension, which alters the wire's resonance frequency. For instance, when the pressure rises, the element increases the tension in the wire, raising the resonance frequency of the wire. To find deviations, a digital counter circuit is employed. This kind of transducer can be utilised for low differential pressure applications as well as absolute and relative pressure sensing since it can detect this frequency change with a certain level of accuracy.

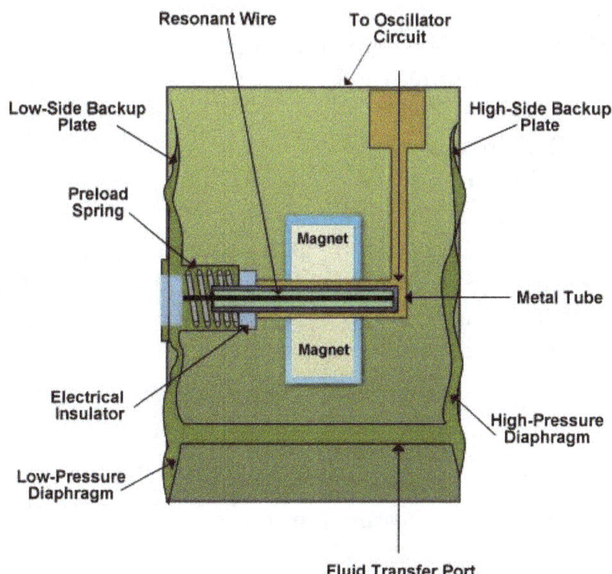

Resonant wire pressure transducer

Exercise

1(a) Discuss the principles of pressure measurement using the following devices:
 i. Manometer
 ii. Bellows
 iii. Bourdon tubes
 iv. Diaphragms

(b) States 3 merits and 2 drawbacks of using the devices
 v. Manometer

vi.   Bellows
vii.   Bourdon tubes
viii.   Diaphragms

2(a) What do you understand by the term "pressure" in engineering measurements.

(b) Explain the following pressure types
i.   Atmospheric
ii.   Gauge
iii.   Differential
iv.   Absolute
v.   Vacuum

3. What relation is there between Q2b (i), (ii) and (iv)

4. Explain briefly the following methods of electrical pressure measurement systems:
1. Strain gauge pressure transducer
2. Piezoelectric pressure transducer
3. Capacitance pressure transducer
4. Inductive pressure transducer
5. Potentiometric pressure transducer
6. Resonant wire pressure transducer

CHAPTER SIX

FLOW MEASUREMENT

6.1   Introduction

Flow measurement is very important in many fields of engineering, especially process control. There are two types of flow measurement namely:

- i)   rate of flow
- ii) total flow

Rate of flow refers to the amount of fluid that flows past a given point at any instant of time.

Total flow is the amount of fluid that flows past a given point at a definite period of time.

Flow rate may be velocity of flow in m/s or volumetric flow rate in m^3/s. The type of fluid and its properties are the major factors which dictate the method of measurement most suitable for a purpose.

6.2   Type of flow

There are two types of fluid flow namely: i) Laminar flow and ii) Turbulent flow

6.2.1   Laminar Flow

Laminar flow is the term used to describe fluid flow that has a low average velocity. The fluid moves in layers, with the fastest layers going toward the center and the slowest layers moving toward the stream's outer edges. This means that the different components of the stream do not appear to have mixed together very much. Here's an illustration of this:

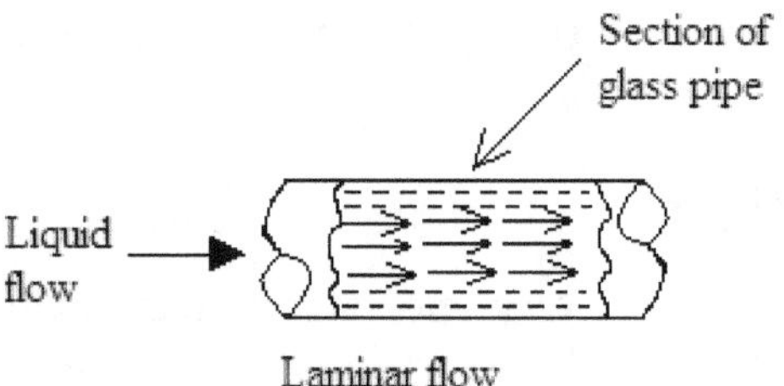

6.2.2   Turbulent Flow

When the velocity of a fluid flow is high, the average velocity across the fluid becomes more uniform making the layer of fluid flow disappearing. The elements of the fluid appear to mix chaotically in the fluid. This is referred to as turbulent flow. The following figure illustrates this.

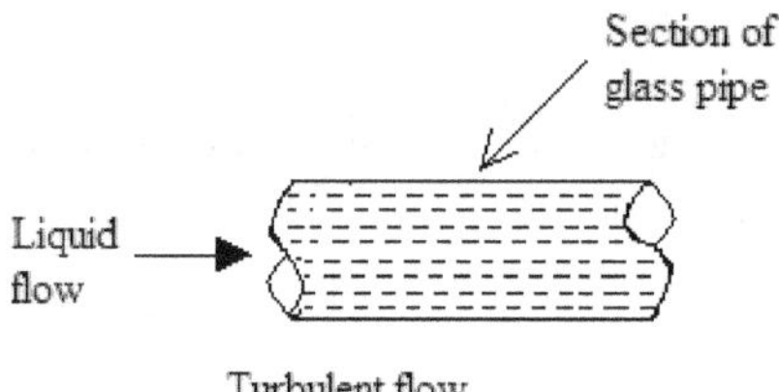

6.3   Compressible and Incompressible fluids

The density of some fluids can easily change with pressure. Such fluids that change their density with application of little pressure are referred to as compressible fluids. On the other hand, a very large pressure is required to achieve a small change in density of other fluids. However, incompressibility does not always imply constant density. For instance, adding salt to water alters the water's density without affecting its volume.

Accordingly, a fluid's compressibility is determined by how much its volume (V) changes when its pressure changes. The bulk modulus of elasticity, E, is the number used to quantify this phenomena. It is expressed as follows:

$$E = -\frac{\Delta p}{(\Delta V)/V}$$

Example: If water has the value 2179 MPa bulk modulus of elasticity, find the pressure that needs to be applied in order to change the volume of water by 1%.

Solution:

$$1\% \text{ change in volume means } \frac{\Delta V}{V} = -0.09$$

$$\therefore E = -\frac{\Delta p}{(\Delta V)/V}$$

$$\Rightarrow \Delta p = -E\,\frac{(\Delta V)}{V}$$

$$= -2179\, x - 0.01$$

$$= 21.79 MPa$$

6.4   Reynolds Number

In a turbulent flow, the average velocity of a specific stream X-section is referred to as the fluid's velocity. It is given as:

$$V = \frac{rate\ of\ flow\ in\ m/s}{area\ of\ pipe\ in\ m^2} = \frac{Q}{A}$$

A term that indicate the nature of fluid flow is the Reynolds number, N_R given as

$$N_R = \frac{vD\rho}{\mu}$$

where v = velocity of flow

D = diameter of pipe

ρ = fluid density

μ = dynamic viscosity of the fluid.

The flow type can be identified as laminar or turbulent using the Reynolds number. When the Reynolds number is less than 2000, the flow is typically laminar. The flow becomes completely turbulent for N_R exceeding 10^5.

Example: Determine the type of flow in glycerine at 25°C if it flows in a pipe with a 150mm inside diameter. The average velocity of flow is 3.6m/s; $\mu = 9.6 \times 10^{-1}$ Pa.s and $\rho = 1258 kg/m^3$.

Solution

$$N_R = \frac{rD\rho}{\mu}$$

$$= \frac{3.6 \times 0.15 \times 1258}{9.6 \times 10^{-1}} = 708.$$

Since $N_R = 708$ which is less than 2000, the flow is laminar.

6.5   Fluid Viscosity

A fluid experiences shear stress during motion, the amount of which is determined by the fluid's viscosity. Shear stress, defined as

$$\tau = \mu \frac{\Delta V}{\Delta y},$$

is the force needed to move one unit area layer of a substance over another.

Where $\mu = $ *the dynamic viscosity*

$$\frac{\Delta V}{\Delta y} = velocity\ gradient$$

$$\therefore \ \mu = \ \tau\left(\frac{\Delta V}{\Delta y}\right) in\ Pa.\,s\ or\ Ns/m^2\ or\ kg/ms$$

Since μ and ρ are both fluid properties, v is also a property of the fluid and has a unit of m²/s. The kinematic viscosity, or nu in Greek, is defined as:

$$v = \mu/\rho.$$

This definition applies to many computations in fluid mechanics that involve the ratio of the dynamic viscosity to the fluid density.

6.6   Bernoulli's Equation

The law of conservation of energy states that energy can only be transformed from one form into another and cannot be created or destroyed. This includes fluid flow in a pipe where there are changes in pressure, velocity, and elevation as shown here. Bernoulli's theorem derives from this law. According to Bernoulli's theorem, there are no energy losses between points 1 and 2, even though there are changes in elevation head, pressure head, and velocity between the two points. In other words, the total heads remain constant between the two points.

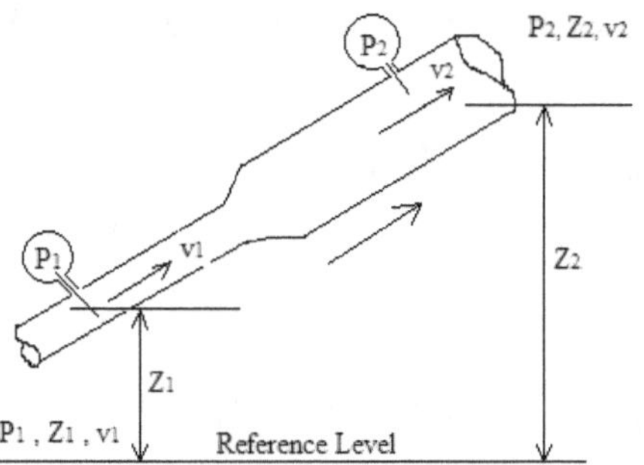

When examining pipe flow problems, three types of energy are always taken into account. For instance, consider the fluid flow at point 1 in the above figure. The following types of energy are present in the fluid element:

1. Potential energy, which is provided as a result of its elevation, given as:

$$PE = WZ$$

where w =weight of the element.

2. Kinetic energy is the energy that a fluid possesses as a result of its motion. It is given as follows:

$$KE = \frac{wv^2}{2g}$$

3. Flow energy, also known as pressure energy or flow work, is the quantity of effort required to move against pressure P. It is represented as follows:

$$FE = \frac{wp}{\gamma}$$

Derivation

Force acting across a section of fluid is $F = pA$

where p = pressure at a particular section

and A = area of the section.

The force propels the element across the section by a distance L equal to the element's length. As a result, the work done is:

$$work = pAL = pV$$

where V = volume of the element.

The weight of the element w is $w = \gamma V$

where $\gamma = specific\ weight\ of\ the\ fluid$

$$\therefore V = \frac{w}{\gamma}$$

$$\Rightarrow work = pV$$

$$= \frac{pw}{\gamma} \text{ as above.}$$

Now the total sum of the energy possessed by the fluid element is

$$E = FE + PE + KE$$

$$= \frac{wp}{\gamma} + wz + \frac{wv^2}{2g} \quad N.m.$$

At section 1, the total energy is

$$E_1 = \frac{wp_1}{\gamma} + wz_1 + \frac{wv_1{}^2}{2g}$$

and at section 2:

$$E_2 = \frac{wp_2}{\gamma} + wz_2 + \frac{wv_2{}^2}{2g}.$$

If no energy is lost between sections 1 and 2, then $E_1 = E_2$

$$\Rightarrow \frac{wP_1}{\gamma} + wz_1 + \frac{wv_1{}^2}{2g} = \frac{wp_2}{\gamma} + wz_2 + \frac{wv_2{}^2}{2g}$$

$$\Rightarrow \frac{P_1}{\gamma} + z_1 + \frac{v_1{}^2}{2g} = \frac{p_2}{\gamma} + z_2 + \frac{v_2{}^2}{2g}.$$

This is referred to as Bernoulli's equation.

Example

The water in the preceding illustration is flowing from Section 1 to Section 2 at 10°C. Section 1 has a diameter of 26 mm, a gauge pressure of 345 kPa, and a flow velocity of 3.0 m/s. Section 2 is located 2.0 meters above Section 1 and has a diameter of 50 mm. Determine the pressure, p_2, assuming that the system experiences no energy losses.

Solution

$$D_1 = 25mm \qquad\qquad V_1 = 3\ m/s \qquad\qquad Z_2 - Z_1 = 2m$$

$$D_2 = 50mm \qquad\qquad p_1 = 345\ \text{kPa (gauge)} \qquad p_2 = ?$$

Now,

$$\frac{P_1}{\gamma} + z_1 + \frac{v_1{}^2}{2g} = \frac{p_2}{\gamma} + z_2 + \frac{v_2{}^2}{2g}$$

$$\frac{P_2}{\gamma} = \frac{P_1}{\gamma} + z_1 + \frac{v_1{}^2}{2g} - z_2 - \frac{v_2{}^2}{2g}$$

$$p_2 = \gamma \left(\frac{P_1}{\gamma} + z_1 + \frac{v_1^2}{2g} - z_2 - \frac{v_2^2}{2g} \right)$$

$$= P_1 + \gamma \left(Z_1 - Z_2 + \frac{v_1^2 - v_2^2}{2g} \right)$$

$$g = \text{acceleration due to gravity} = 9.81\ m/s^2$$

$$\gamma = 9.81\ kN/m^3$$

Using continuity equation, V_2 is determined thus:

$$A_1 V_1 = A_2 V_2$$

$$V_2 = V_1 \frac{A_1}{A_2}$$

$$A_1 = \frac{\pi D_1^2}{4} = \pi \frac{(50^2 m)}{4} = 1963 mm^2$$

$$\therefore V_2 = 3 \, x \, \frac{491}{1963} = 0.75 m/s$$

$$\therefore P_2 = 345 + 9.81 \left(-2 + \frac{(3^2 - 0.75^2)}{2 \, x \, 9.81}\right)$$

Note that $Z_1 - Z_2 = -2$m since Z_2 is 2m greater than Z_1

$$p_2 = 345 + 9.81 \, (-2 + 0.43)$$

$$= 345 - 15.4$$

$p_2 = 329.6 \, kPa$ (gauge) for it is computed relative to P_1 which

is also a gauge pressure.

6.7 Differential Pressure Measurement

According to the differential pressure measurement principle, differential pressure is the difference between two applied pressures. It is sometimes denoted as DP or Δp. For instance, the differential pressure is 85 psi (195 pa – 110 pa) if the pressure at point A is 195 psi and the pressure at point B is 110 psi.

A venturi tube, flow nozzle, or orifice plate can all be used to monitor differential pressure. A known-dimension restriction is typically inserted into the pipeline when using differential type flow meters. When the flow velocity increases, a head loss or pressure drop happens at the constriction. It is proportional to the square of the flow rate to measure this pressure drop. Operating on the same principle, all three of the above-mentioned devices offer a straightforward method of creating a pressure drop.

6.8   Advantages of Differential Flowmeters

The following are the advantages of differential pressure flowmeters.

- They can be used with liquids, gasses, and vapours in any situations.
- Suitable for harsh conditions like viscosity.
- They enable calculations for atypical circumstances.
- Fit for high pressures and temperatures.
- Variations in range possible.
- They provide a little dip in nozzle pressure.

6.9   Disadvantages of Differential Flowmeters

Listed below are the disadvantages of the differential flowmeters.

- There is a square root relationship between flow rate and differential pressure, and therefore a small span.
- They are affected by changes in pressure and density.
- There is pressure drop across the measuring orifice.
- The sharpness of the edges of the measuring orifice must be ensured, so that there are no solids or impurities present.
- Very long inlet and outlet paths
- Requires expensive installation, differential pressure lines, fittings, and sensors.
- Their use requires additional installation and maintenance experience.

6.10   The Orifice Plate

The most popular and basic flow rate sensing component is this one. The thin, round metal plate that makes up the orifice plate has a hole in it. It is contained in the pipe line in the space between two orifice flanges. Concentric, eccentric, or segmental orifices are possible. The orifice plate is made of a variety of materials, such as brass, phosphor bronze, Monel metal, stainless steel, and steel. The flow of a pipe will abruptly compress as it gets closer to the orifice and then abruptly expand again to the full pipe diameter when the orifice plate is inserted within the pipe. This causes the flow velocity to increase quickly, which lowers the pressure downstream of the orifice.

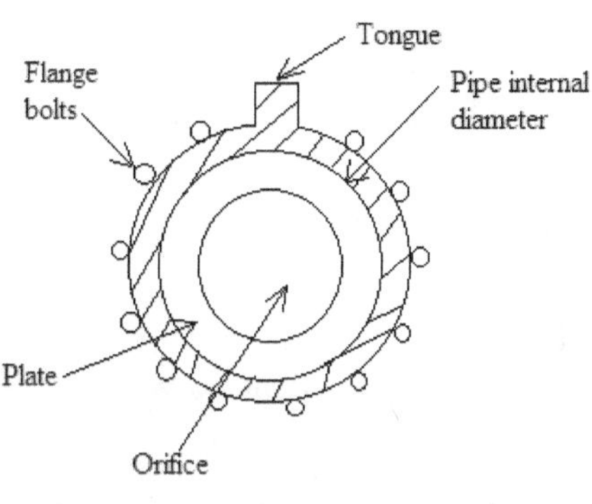

Concentric type of orifice plate

The primary factor to be taken into account when choosing an orifice plate is the ratio of its opening (d) to the internal diameter (D) of the pipe (d/D). The pressure loss increases too much if this ratio is too little, and it becomes too little to notice if it is too great.

The orifice plate is very simple to install and replace, and it is reasonably priced. It does not, however, have a decent recovery rate and is not very accurate. It cannot be used with high $\frac{d}{D}$ ratio.

When a measuring device like an orifice plate, venture tube, or flow nozzle reduces the area of flow, the flow changes along with the velocity of flow. The flow rate is given as

$$Q = CEBA_2 \sqrt{\frac{2\Delta P}{P_1}} \; m^3/s$$

Where C = coefficient of discharge with equals to the ratio of actual flow rate and the theoretical flow rate.

E = velocity of approach factor which equals

$$E = \sqrt{1 - \left(\frac{A_1}{A_2}\right)^2}$$

B = expansion ratio

A_1 = area at the smallest part of pipe (m^2)

A_2 = area at point x (m^2) ie widest diameter

ρ = density of fluid (kg/m^3)

ΔP =
pressure drop between the point of smallest and greatest diameter

6.11   The Venturi Tube

Sections of gun metal, stainless steel, or cast iron are arranged in a circular, square, or rectangular shape to form a venturi tube. In its most basic form, it is a specially formed piece of pipe. As shown here, the device resembles two funnels linked at their little entrance. For extended pipe lengths, the venturi tube is utilised. Although it is significantly more expensive and more complex to install, it is more accurate than orifice plates.

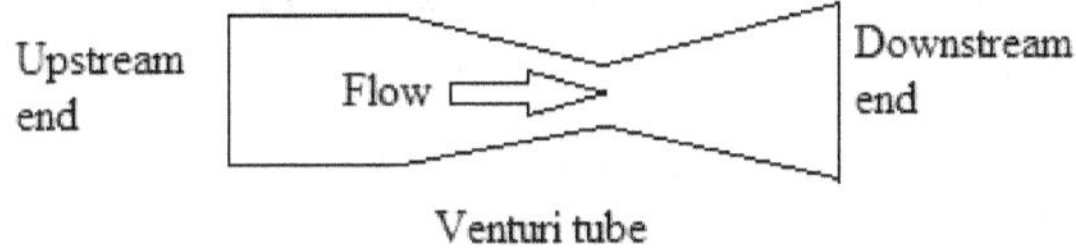

Venturi tube

There is not much pressure drop through the venturi tube because of its outstanding pressure recovery. The device effectively resists wear and does not impede abrasive sediment.

6.12   Measuring Pressure with Flow Nozzles

The flow nozzle is designed to function as the venturi tube's entering half. Where a pressure differential is produced during flow, this is the main flow rate detecting element. Because it combines the low losses of the venturi tube with the simplicity of the orifice plate, this device is popular in many applications. It is a little less accurate and does not offer as excellent pressure recovery as the venturi tube, though less complicated and more affordable.

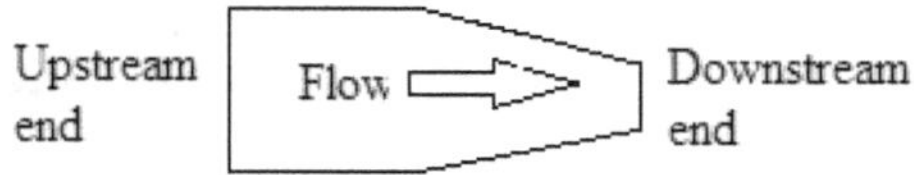

6.13   Open Channel Measurements

Weirs and flumes are often used for open channel flow measurement.

6.13.1 Weirs

A weir is essentially a dam with either a rectangular, trapezoidal or V-notch through which liquid flows. A weir is employed where sufficient head or fall can be obtained. When it is positioned over an open fluid stream, the fluid is forced to ascend up the notch as the flow increases.

6.13.2 Flume

A constructed structure known as a flume is inserted into an open fluid stream to force the fluid to rise within it as the flow rate increases.

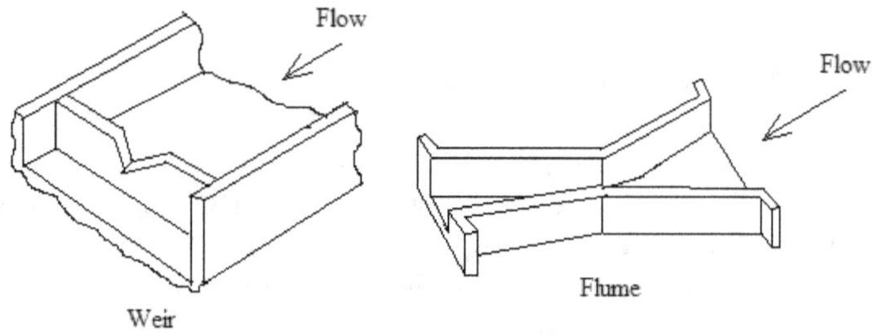

Example

To measure the volumetric flow, a 60 mm-thick venturi tube is inserted into a 100 mm-diameter water pipe. The water in the tube has a density of 10^3kg/m^3 and viscosity of 10^{-3} Ns/m^2. The volumetric flow rate through the tube is 0.08 m^3/s.

a) Determine the Reynolds number for these conditions
b) The coefficient of discharge is 0.99 and the expansion ratio is unity, determine the upstream to throat different pressure

Solution

a) Area of throat $= \dfrac{\pi d^2}{4}$

$$= \frac{\pi}{4} \times (60 \times 10^{-3})^2$$
$$= 2827 \times 10^{-6} \text{m}^2$$

Velocity through throat $v = \dfrac{0.08}{2827 \, x \, 10^{-6}}$
$$= 28.3 m/s$$

Reynold's no, $N_R = \dfrac{vdp}{\mu}$

$$= \frac{28.3 \, x \, 60 \, x \, 10^{-3} \, x \, 10^3}{10^{-3}}$$
$$= 1.7 \, x \, 10^6.$$

b) $E = \sqrt{1 - \left(\dfrac{A_1}{A_2}\right)^2} = \sqrt{1 - \left(\dfrac{60}{100}\right)^2} = 0.8$

$$Q = CEBAy\sqrt{\frac{2\Delta P}{\rho}}$$

$$Q = 0.99 \, x \, 0.8 \, x \, 1 \, x \frac{\pi}{4}(60x \, 10^{-3})^2 \sqrt{\frac{2\Delta p}{1 \, x \, 10^3}} = 0.08$$

$$\therefore \Delta p = 638.1 \, KN/m^2.$$

Exercise

1. Explain the principle of operation of the following differential pressure measuring devices:
    a. Orifice plate
    b. Flow nozzle
    c. Venturi tube
2. Describe the following open channel pressure measurement devices:
    a. weirs
    b. Flume
3. In a hydraulic system, water having specific weight of 9.81 kN/m^3 flows in a channel at a rate of 0.37 m^3/s as shown in the figure. Calculate the outlet pressure if the pressure at point A is 66.2 kPa.

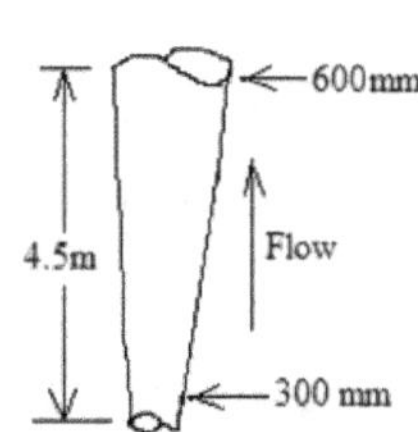

4. Determine the pressure in the pipe immediately in front of the nozzle in the figure to generate a jet velocity of 22.86 m/s. At 80 °C, the fluid is water with a specific weight of 9.53 kN/m3.

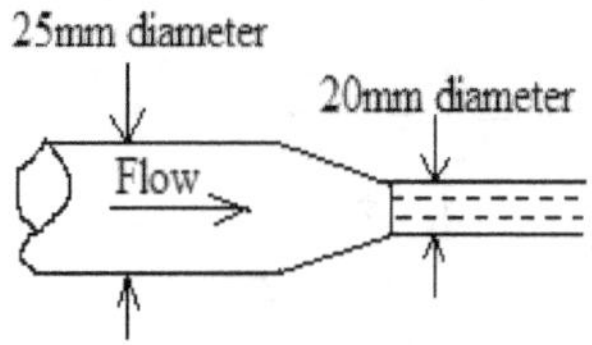

5. For the system shown in the accompany figure, calculate:
    (a) the water's flow rate out of the nozzle and
    (b) the pressure at point A.

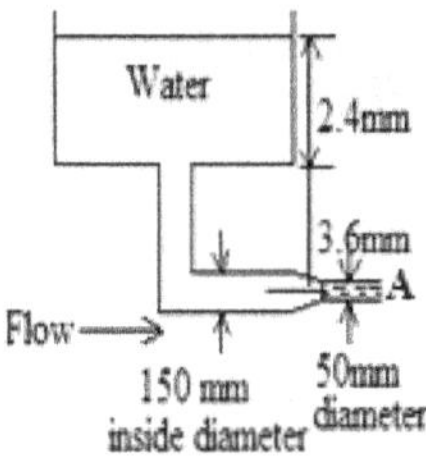

6. Calculate the percentage change in the volume of the water in a certain hydraulic system operating at 20MPa. $E_{H_2O} = 2179MPa$.

7. A hydraulic system running at 20.7 mPa in a cylinder with an inner diameter of 25 mm and a length of 305 mm saw a 1% volume change. With this volume change, calculate the axial distance the piston would travel. The oil that is used has a bulk modulus of elasticity of 1303 mPa.

7. In a copper tube of 25.27 mm in diameter and $55.017 \times 10^{-4} m^2$ in area, water flowing at 70°C has a flow rate of 285 liters per minute. Determine whether the flow is laminar or turbulent.

Answer: $N_R = 5.82 \times 10^5$, turbulent flow.

CHAPTER SEVEN

LEVEL MEASUREMENT

7.1   Introduction

Many fluid flow systems include bulk storage tanks as essential components, and it is frequently necessary to keep an eye on the fluid level in these tanks. Level readings can initiate automatic level control and are usually sent to distant monitors or central control stations. Monitoring and quantitatively measuring the liquid content in tanks, vessels, reservoirs, the height of a column in open-channel streams, and other similar situations in industrial processes are common uses for liquid level measurement.

Some common methods of accomplishing the liquid level measurement include:

i)	Resistive method
ii)	Float method
iii)	Force balanced diaphragm system
iv)	Bubbler or purge system
v)	Capacitive method
vi)	Ultrasonic method
vii)	Radiation method
viii)	Eddy current method

7.2   Resistive Method

For detecting liquid level, the most basic electrical method is the resistive method. It is also referred to as the contact point method. The method is illustrated as follows. A mercury column is operated by liquid column. A number of resistances of suitable values are placed at various levels of liquid, as shown in the figure. As the level of liquid rises, the mercury level also rises and shorts the successive resistance and so the resistance R decrease or current through the indicator increases. Resistances $r_1, r_2 \ldots r_n$ are so chosen that 1/R is a linear function of liquid level, thus. The liquid level is read directly from the indicator.

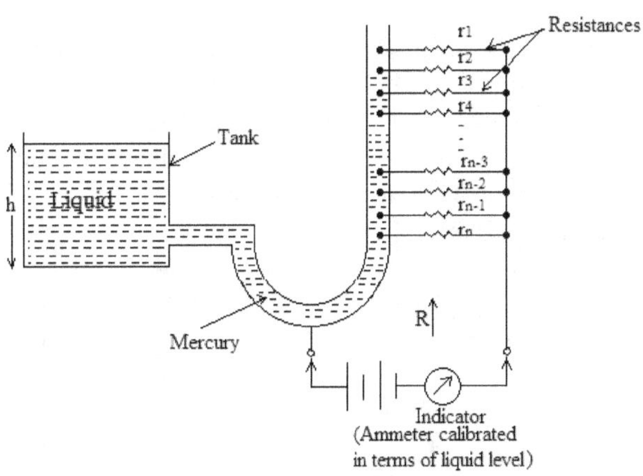

Advantages

i) The technique is straightforward and able to provide a direct indication of liquid level.

ii) The system operates on low voltage and so there is no danger to the operator and no possibility of arcing at the contact points.

iii) The signal can be transmitted to any desired point.

iv) Fairly continuous record of liquid level can be had by adding more resistances, with separate signal output for each contact resistance.

Disadvantages

i) This measuring system's primary flaw is that the liquid could cause a significant mistake.

ii)   Another drawback is that the measuring system has to be calibrated differently for different liquids.

7.3   Float Method of Level Measurement

A float rises or falls in response to changes in the fluid level due to the buoyant force acting upon it. A switch or signal can be activated by the float and sent to distant devices. The float is mechanically coupled to some suitable displacement transducer, such as simple potentiometer or LVDT, for continuous indication and recording. This is illustrated below.

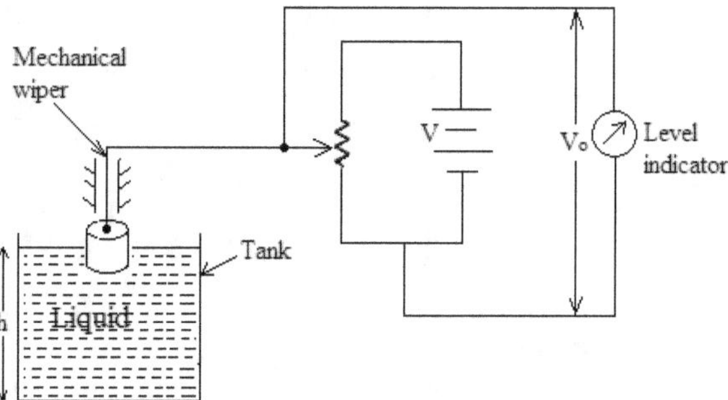

As the liquid level varies, the float moves up or down thereby causing the potentiometer wiper to change position as well. Thus voltage reading on the indicator varies according to changes in liquid level. The output voltage V_o is proportional to the liquid level h.

The method's advantage is that the output terminals of the potential divider can be moved to distant locations for control and display because the output voltage is proportionate to the liquid level h. The drawback of the float system is liquid level measurement is that any changes affect the accuracy.

7.4   Force balance diaphragm method

In this method, the technique of measuring differential pressure formed by hydraulic level and liquid column are used as shown below.

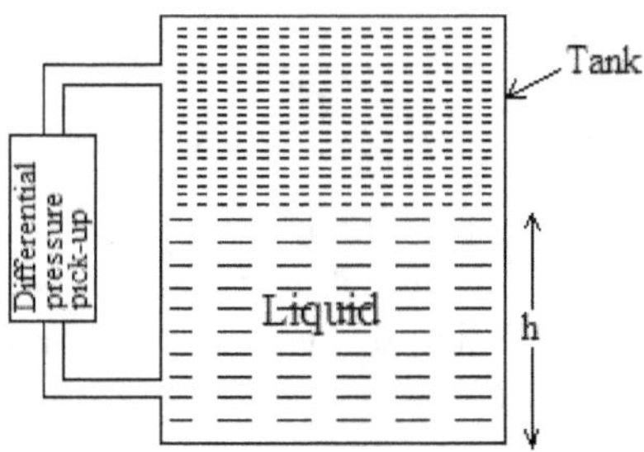

The actual height of the liquid column determined, knowing the liquid density and making corrections for the variations in temperature and mass density. The sensor used for the measurement is the differential pressure cell, using special corrosion materials for the diaphragms. The method of liquid level gives high accuracy and greater resolution.
Its demerit lies in the high cost of construction in compared to other methods.

7.5   Bubbler or Purge system

In the bubbler or purge system of measurement of liquid level, a pipe is inserted into the tank within a few cm of the bottom and air or some other fluid is pumped through the pipe as shown in the accompany figure.

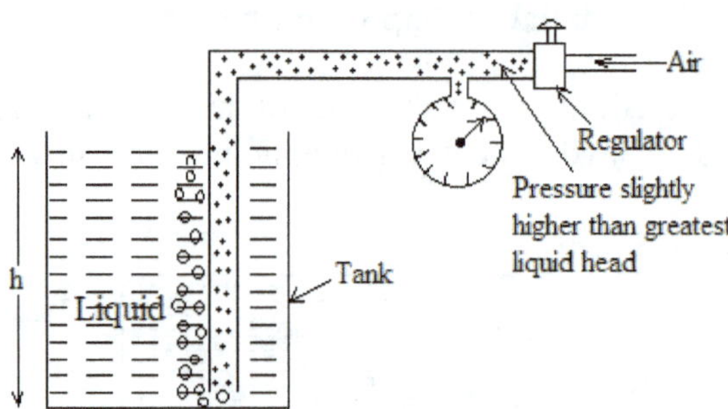

When it bubbles, the air pressure in the pipe is equal to the hydraulic head of the liquid and this can be measured with standard differential pressure transducer. The liquid level is then computed knowing the mass density of the liquid.

7.6   Capacitive Method of level Measurement

Under this technology, a sensor receives a high-frequency AC electrical signal, and the probe's depth submergence and capacitance determine how much current flows through the device.

The liquid in the tank serves as the capacitor's dielectric, and the capacitive device is essentially an electrical probe of the concentric variety that is inserted into a tank. The liquid level is directly indicated by electrically measuring the capacitance of the probe, which changes in response to changes in the fluid level. An ac bridge system or a high frequency oscillator linked with a tank circuit can be used to measure capacitance precisely as a change in frequency. The following is a typical configuration:

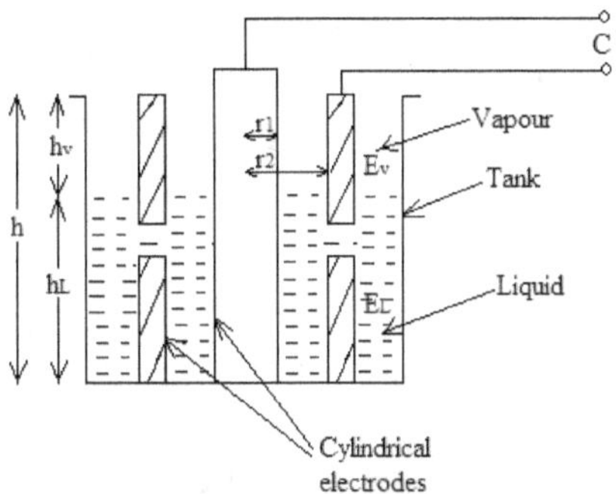

The electrodes are arranged in the form of a pair of concentric cylinders in the tank with the holes in the outer cylinder at the lower and to allow passage of liquid. As the level of liquid rises, the capacitance of the capacitor between the electrode pair increase because the dielectric constant of vapours (ε_v) is very small in comparison to that of liquid(ε_l). The capacitance of the arrangement is given as:

$$C = 2\pi\varepsilon_o \ \frac{\varepsilon_l h_l + \varepsilon_v h_v}{\log e \frac{r_2}{r_1}}$$

Example
The capacitance between the electrodes in a capacitive level measurement method is found to be 0.00567μF. Assuming that the height of vapour is 600 cm, find the height of liquid (castor oil) in the tank if the diameters of the inner and outer electrodes are 20 cm and 125 cm, respectively. ε_l (castor oil) is 4.7, ε_v is 221.3, and ε_0 is 8.854 x 10^{-12} F/m.

Solution

$$C = 2\pi\varepsilon_o \ \frac{\varepsilon_l h_l + \varepsilon_v h_v}{\log e \frac{r_2}{r_1}}$$

We re-arrange the formula to make the unknown the subject of the relation thus:

$$h_l = \frac{1}{\varepsilon_l}\left(\frac{C \, \log e \, \frac{r_2}{r_1}}{2\pi\varepsilon_o} - \varepsilon_v h_v\right)$$

$$h_l = \frac{1}{4.7}\left(\frac{0.00567x10^{-6}}{2\pi x 8.854 \text{ x } 10^{-12}} \log e \, \frac{125}{20} - 1.3x0.6\right)$$

$$= \frac{54.11}{4.7}$$

$$= 11.51m.$$

7.7   Ultrasonic Method of Liquid Level Measurement

This method measures solids or liquid level by mounting an ultrasonic transmitter on the top of the tank, as illustrated.

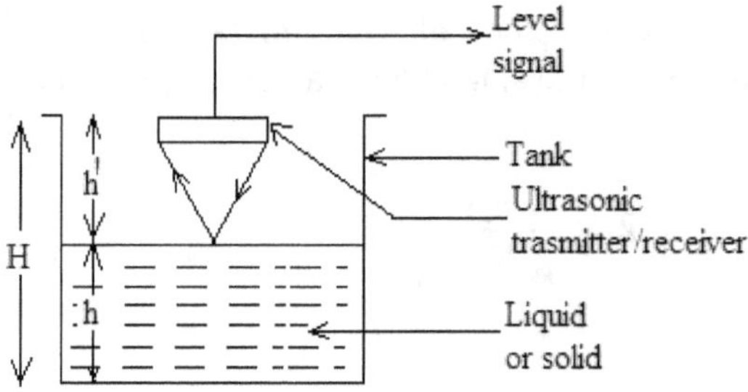

The transmitter projects the beam downward, and the surface of the solid or liquid inside the tank reflects it back. The receiver receives the beam. The beam's travel distance can be calculated from the time it takes. As a result, the distance "h" between the ultrasonic set and the surface of the tank's contents determines the time "t" that passes between sending and receiving a pressure pulse. i.e.

$$T \, \alpha \, h' \, \alpha \, (H - h).$$

7.8   Liquid Level Measurement with Gamma Rays

Using this method, a gamma-ray source is positioned at the bottom of the tank, and a gamma-ray sensor is placed outside the tank near the top. The liquid level affects how intense the rays are. The sensor will be exposed to the most radiation if the tank is empty. In addition, some of the rays will be absorbed by the liquid in the tank, resulting in less radiation reaching the sensor. As a result, the sensor's output and the liquid level are inversely related.

The sensor's output is in the form of pulses, which a counter can count. Consequently, it is possible to immediately calibrate the counter using the liquid level.

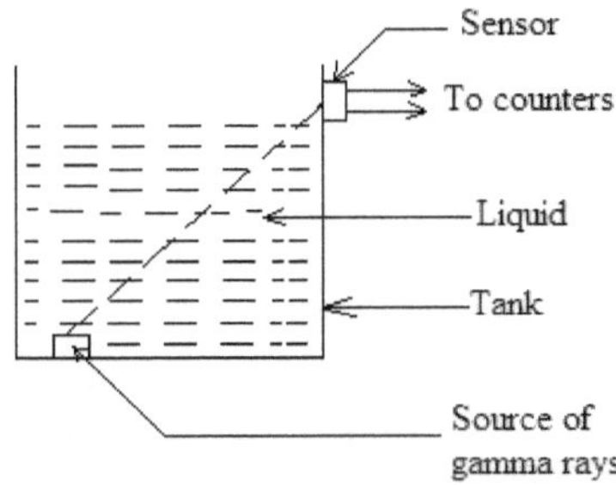

Exercise

Explain the operating principles of the following methods of liquid level measurement:

- i)   Resistive method
- ii)   Float method
- iii)   Force balanced diaphragm system
- iv)   Bubbler or purge system
- v)   Capacitive method
- vi)   Ultrasonic method
- vii)   Radiation method

viii)   Eddy current method

ix)   In a capacitive method of level measurement, it is found that the capacitance between the electrodes is $0.0067\mu F$. If the diameters of the inner and outer electrodes are 15 cm and 120 cm respectively, calculate the height of liquid (castor oil) in the tank if the height of vapour is 60 cm. (ε_0 is 8.854 x 10^{-12} F/m ε_v is 1.3 and ε_l (castor oil) is 4.7).

CHAPTER EIGHT

INSTALLATION OF INSTRUMENTS

8.1   Introduction

The accuracy and dependability of the measuring devices in an installation have a significant impact on the quality of the job that is completed. By providing the conditions for routinely inspecting or calibrating the equipment at its operating area, a significantly higher reliability margin is attained for the apparatus. Notably, this kind of verification needs to be carried out in actual operational environments, which are very different from typical ones.

In case a monitored measuring instrument is situated outside the installation, group comparisons can be made by adding a set of standard measuring instruments to it and applying the same measured quantity to its input as if it were a measuring instrument inside the installation. It is feasible to verify that an operational instrument is in good metrological working order through this type of monitoring.

8.2   Industrial Safety

Safety of lives and property is of great importance in industrial systems and premises. There is the need to protect against equipment damage, not to mention damage to personnel and potential hazard to the environment. Therefore, it is important to install and operate correctly, many safety devices in the industries.

Many hazards are present in the industries that must be avoided. These hazards vary from fire, rupturing devices, flammable liquids or gases, etc. Each of these industrial hazards has developed safety measures to avoid a breakout of disaster in industries. When any of these hazards are present, it is necessary that a risk assessment is carried and necessary precautions and safety measures put in place. Such safety measures will be tailored to the immediately industrial need in the larger environment as a whole. For instance, if flammable gases present a known or possible risk at work, it's critical to take he following steps:

- All flammable gases present at the worksite should be listed, along with their risk areas and uses.

- Employees should be trained on the characteristics and potential issues of each gas, including potential ignition sources.

- Each gas should also undergo a risk assessment procedure.

- Finally, a plan should be developed to maximize the safety of both personnel and facilities.

The following safety devices necessary for the installation of instrument and control devices shall be discussed:

- Pressure relief valves
- Rupture Discs
- Flame arrestor
- Flammable detector.

8.3   Pressure relief valves

Devices that detect and prevent overpressure safeguard systems and equipment. When used for vapour relief, a pressure safety valve (PSV) is sometimes referred to as a Safety Relief Valve (SRV), and when used for liquid relief, it is known as a Pressure Relief Valve (PRV) (not to be confused with a Pressure Reducing Valve). When the pressure rises over the set point, these valves have the ability to open, and they close when the pressure falls below the set threshold. In addition to protecting persons, property, and the environment from harm, PRVs are the first line of defense against equipment damage. It is crucial to install and use it correctly as a result.

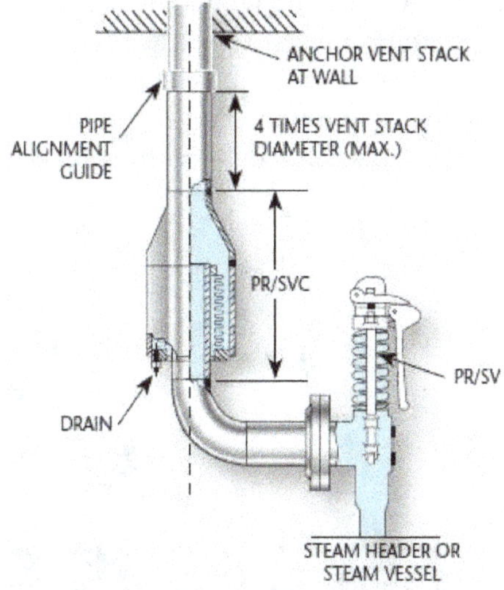

When using PRVs, a pressure safety mechanism is required to release and regulate excess pressure and prevent the vessel from rupturing if the system pressure rises above the rated pressure limit of 125 Pa for a pressurized tank. Pressure safety devices are often pre-fitted in vessels with a pressure rating. The guidelines for installing PRVs are as follows.Operate the valve so that it does not get as well close to the setpoint. It may cause bubbling or spillage. Once this begins it tends to get worse.

- Do not test valves habitually. Lifting the test lever as well regularly can trap earth and other flotsam and jetsam within the situate and cause the valve to spill.

- Regular testing is recommended as part of a preventative maintenance program, but should be no more than once a year.

8.4   Rupture Discs

A rupture disc is a safety device with a set breaking point that reacts to a certain pressure and is used to release pressure in a number of applications. It is also known as a pressure safety disc, burst disc, bursting disk, or burst diaphragm. In the process, they guard against overpressure and vacuums that could harm people, the environment, and machinery.

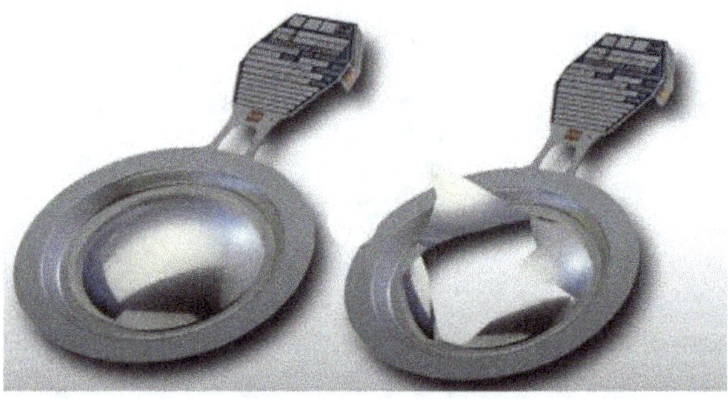

The purpose of rupture discs, which are non-resealing pressure relief devices, is to shield process piping systems, equipment, and pressure vessels from harmful vacuum situations or overpressurization. The two most often utilised pressure protection devices in industrial plants are rupture discs and pressure relief valves. A disposable diaphragm on a rupture disc bursts at a specific differential pressure—gauge or vacuum. When there is either positive or negative pressure in the process piping system, rupture discs react immediately, but they do not reseal after rupturing. When comparing rupture discs to pressure relief valves, the key benefits are economy, dependability, and leak resistance.

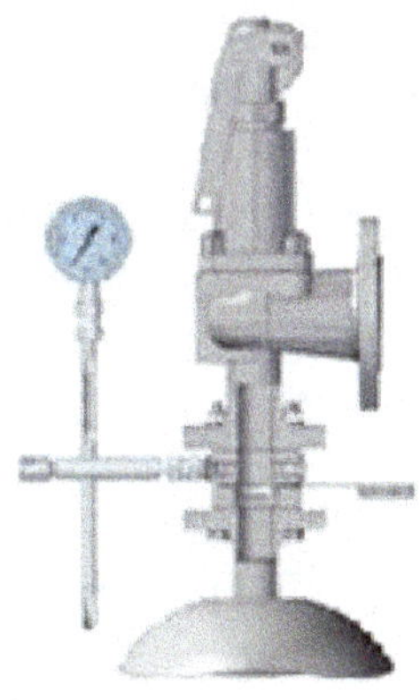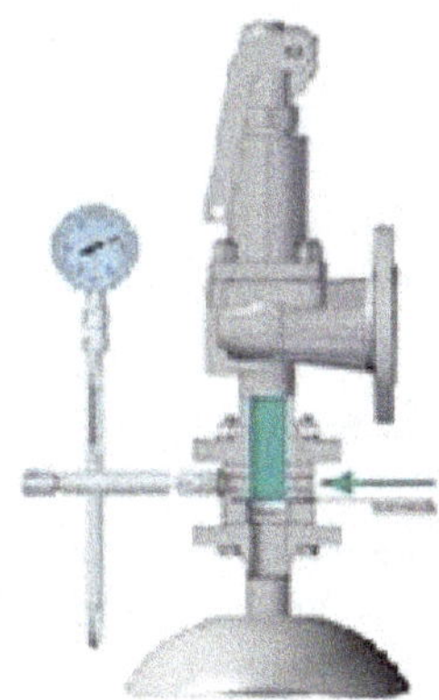

Ruptured discs are installed in two different configurations along with pressure relief valve.

8.4.1   A rupture disc mounted parallel to the pressure relief valve

The rupture disc will activate and burst to manage the pressure in the event of overpressure if the pressure relief valve malfunctions or is unable to relieve the overpressure fast enough.

8.4.2   Rupture disc installed under the pressure relief valve

Generally speaking, pressure relief valves leak more after their first trip. To guarantee an ideal leak-tight seal, a rupture disc can be placed   in   front   of   the   pressure   release   valve.

In process line systems, upstream process burst discs help guarantee the performance and dependability of pressure relief valves for corrosive, sticky, polymerizable, or viscous liquids. For corrosive liquids, this permits the use of less expensive materials for pressure relief valves. Replacing the rupture discs is also less expensive than replacing the pressure relief valve. It is possible to test pressure relief valves without taking them out of the process piping system. This is accomplished by pressurizing the area between the burst disc and the valve lift. Since the rupture disc has nearly twice the back pressure strength, the safety valve examination won't harm it.

8.4.3   Advantages of rupture disks

- Rupture disks react rapidly enough to release the excess pressure; they can be equipped with one or more pressure relief devices to protect the vessels.
- The surplus pressure is released by rupture disks in a timely manner.

- The rupture disc reacts more quickly and simply than other pressure relief devices because it doesn't have any moving parts.

- It is lighter in weight compared with other pressure relief devices.

- It is used to relieve excess pressure in applications involving the handling of gases or liquids..

- Rupture disk is of one-time use and thus no maintenance cost is incurred.

8.4.4   Disadvantages of ruptured disks

- The rupture disk must be replaced in its entirety once it has burst.

- To replace a ruptured disk, the entire system needs to be shut down.

- Highly trained personnel are required for the installation of rupture disk; as much as a small scratch during installation can damage the rupture disk.

- An improper bolt torque during installation also affects the disk burst pressure.

8.5   Flame arrestor and Detectors

Industrial flame arrestors (arrester) are safety devices used to ensure the overall safety of organisation and employees from flammable or combustible gases hazard. They are crucial in many industries, such as mining, refining, petroleum extraction, and manufacturing. There are various technologies of these devices to choose from by any industry – each functioning under different operating principles, and each with different strengths and weaknesses.

Any gas that ignites when combined with oxygen or air is considered combustible. At varying concentrations and temperatures, diferrent combinations of gas and air will burn. Thus, a varying level of protection against the hazard is available. Flammable gases vapours are produced by many industrial processes. These products can burn when mixed with air, and in many instances, violently. The following are examples of such industries:

- flammable materials removal from tanks and pipes in preparation for entry, line breaking, cleaning, or hot work such as welding;
- flammable solvents evaporation in a drying oven;
- spraying, spreading and coating of articles with paint, adhesives or other substances containing flammable solvents;
- flammable gases manufacture;
- flammable liquids manufacture and mixing;

- storage of flammable substances;
- processes of solvent extraction;
- oil or gas combustion;
- combined heat and power plants;
- heat treatment furnaces that use flammable environments;
- charging of batteries.

The flammable gas detector can make a significant difference in how safe these procedures are. When a given gas or vapour concentration is surpassed, they can be utilised to set off alarms. This serves to keep people safe by giving early warning of potential dangers. The detector does not, however, stop leakage. As a result, they cannot take the place of safe industrial processes.

8.6 Instrumentation and Control Symbols

Specific symbols are used in Piping and Instrumentation Diagrams (P&IDs) to illustrate how sensors, valves, and other control system components are connected. These symbols may be seen in the majority of system diagrams, if not all of them, and can stand in for actuators, sensors, and controllers. P&IDs provide more information than a process flow diagram, aside from the parameters (temperature, pressure, and flow values). Process equipment, valves, instruments, and pipe lines are uniquely identified by codes that are attached to them and arranged based on factors such as size, material and fluid contents, connection type (flanged, screwed, etc.), and state (valve - normally closed, normally open). To create a fully functional process, the parameters and control system can be connected using these two diagrams. Because standard notation is a common language used in the industrial sector to discuss plants, engineers should be familiar with it. Standard notation can range from letters to figures.

P&IDs can be manually or digitally manufactured. Popular PC and Mac applications that generate P&IDs are OmniGraffle (Mac) and Microsoft Visio (PC). Similar to other P&IDs, these programs provide relative locations rather than the actual size and location of the sensors, valves, and other equipment. These tools are useful for creating organized P&IDs that are computer-readable and storeable. P&ID templates for various programs can be found below.

Pipeline and instrumentation designs employ a lot of symbols. Here are a few of them that pupils ought to become acquainted with. This is not a comprehensive list, though. No matter how complex a P&ID is, you should be able to recognize a handful of its common symbols if you are familiar with the ones used to create instrumentation diagrams. The first thing one needs to do when coming across a new P&ID is to look over the legend section, which lists all of the symbols used on the P&ID and can help you make sure you are reading the correct equipment. In this manner, the information being provided and what you already know won't clash.

Safety devices necessary for the installation of instrument and control devices

| Pipe | Therma lly insulate d pipe | Jackete d pipe | Cooled or heated pipe |

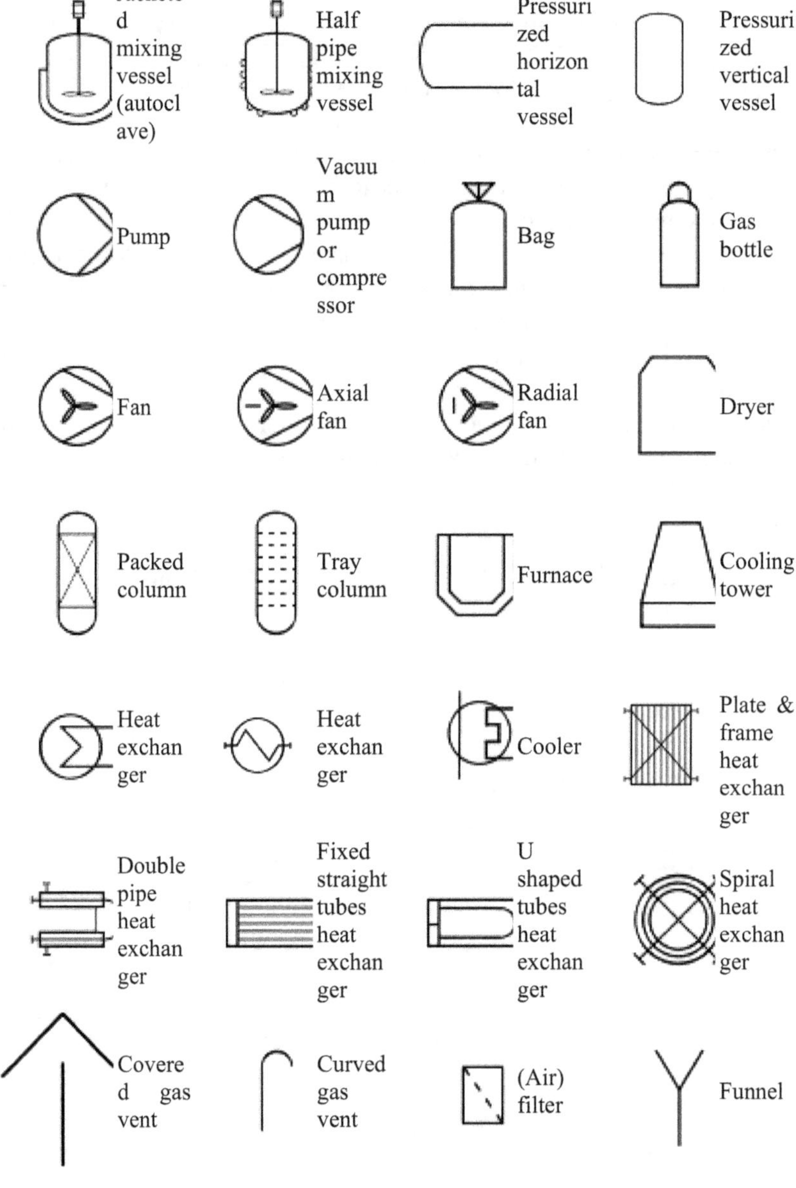

Jacketed mixing vessel (autoclave)
Half pipe mixing vessel
Pressurized horizontal vessel
Pressurized vertical vessel
Pump
Vacuum pump or compressor
Bag
Gas bottle
Fan
Axial fan
Radial fan
Dryer
Packed column
Tray column
Furnace
Cooling tower
Heat exchanger
Heat exchanger
Cooler
Plate & frame heat exchanger
Double pipe heat exchanger
Fixed straight tubes heat exchanger
U shaped tubes heat exchanger
Spiral heat exchanger
Covered gas vent
Curved gas vent
(Air) filter
Funnel

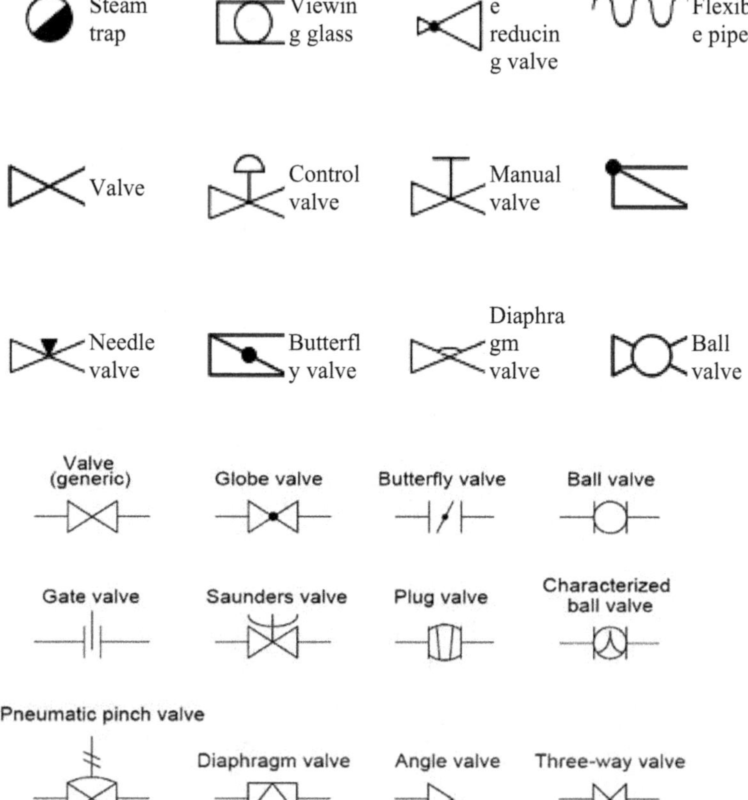

Steam trap
Viewing glass
Pressure reducing valve
Flexible pipe
Valve
Control valve
Manual valve
Needle valve
Butterfly valve
Diaphragm valve
Ball valve
Valve (generic)
Globe valve
Butterfly valve
Ball valve
Gate valve
Saunders valve
Plug valve
Characterized ball valve
Pneumatic pinch valve
Diaphragm valve
Angle valve
Three-way valve

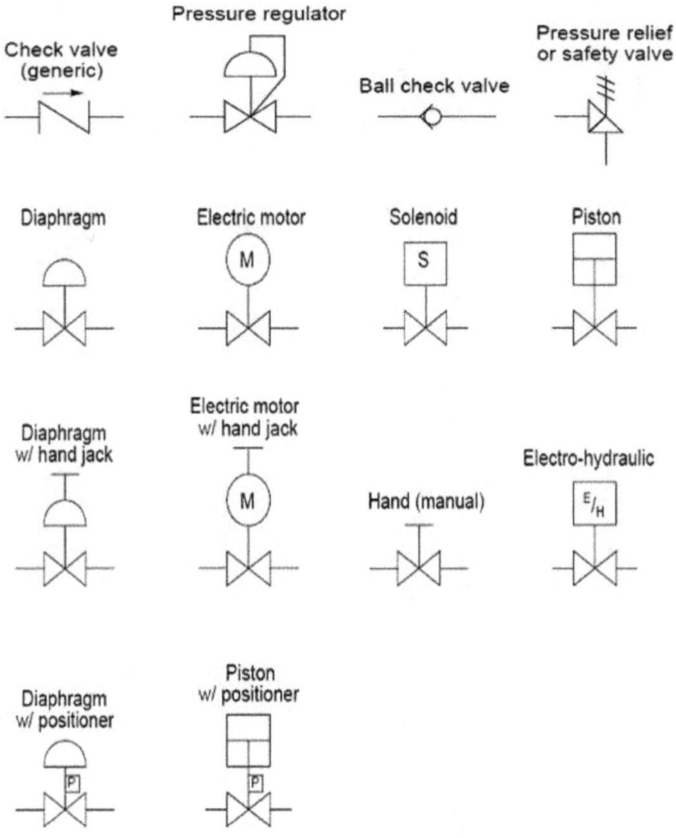

Typically, a sizable control system includes both controllers and instruments. A Piping and Instrumentation Drawing (P&ID), which depicts the devices, their locations, and the method of interconnection, is typically used to describe a system. As previously said, the accompanying graphic displays a typical example of P&ID together with the essential symbols.

The devices are shown as balloons, which are circles with a special tag that is divided into two sections. Two or more letters outlining the function make up the first section. The second part is the device's unique identification number, such as FE127, which stands for flow sensor number 127. To facilitate maintenance, it is a good idea to physically affix the P&ID tag to equipment.

8.7 How to read P&IDs

The complexity of the design influences the level of instrumentation detail. For instance, compared to fully developed piping and instrumentation diagrams (P&IDs), simplified or conceptual designs—also referred to as process flow diagrams—provide less detail. Graphical components, function blocks, connecting lines, abbreviations, and alpha and numeric identification codes are used to form each sign.

Four graphical elements—discrete instruments, shared control/display, computer function, and programmable logic controller—are frequently used in P&IDs. These elements are arranged into three location categories (main location, auxiliary location, and field mounted).

A single horizontal bar placed in the space between any of the four visual elements indicates that the function falls into the major location category. When there is no line, a device or function is placed in the field; an auxiliary location is indicated by a double line. Devices that are concealed beneath panel boards or in other unreachable areas are indicated by a dashed horizontal line. The ISA standard specifies the letter and number combinations that are used in each graphical element. Users assign numbers, and there are variations in schemes. Some organisations employ sequential numbering, while others associate the process line number with the instrument number. Some organisations utilise unusual numbering schemes, sometimes quite unfamiliar.

Measured or initiating variables like analysis (A), flow (F), temperature (T), and so on are defined in the first letter. Functions such as indicator (I), record (R), transmit (T), and so forth are defined by the following alphabets.

8.8 Use of Tag Numbers and Identification Letters

In reading a P&ID diagram, tag numbers are very useful. Physical instruments that stand alone are identified by a tag number surrounded by a circle. Tag numbers are a string of characters and digits that serve to identify a device by identifying what it is controlling, what kind of device it is using, and the P&ID number that corresponds to it. The following points should be noted.

- The parameters that are being measured, tracked, or controlled are indicated by the first letter.

- The sort of device being utilised is indicated by the second letter.

- Letters 3, 4, and 5 change the meaning of the previous letters and further specify the component's function.

As an illustration, "PC" stands for pressure controller and "PIC" for pressure indicator controller. The accompany chart gives examples of how tag number are utilised to represent specific components.

TYPICAL TAG NUMBER

```
TIC 103  - Instrument Identification or Tag Number
T    103  - Loop Identifier
      103  - Loop Number
TIC        - Function Identification
T          - First-letter
IC         - Succeeding-Letters
```

EXPANDED TAG NUMBER

```
10-PAH-5A  - Tag Number
10              - Optional Prefix
       A  - Optional Suffix
```

Note: Hyphens are optional as separators

Following are several combination of letters used to identify various equipment in P&IDs.

- AIC, AI, AT, AY
- FRC, FIC, FC, FY, FE,FQI
- II, IIC, IT
- LIC, LI, LT, LE,
- PIC, PC, PCV, PI, PSH, PIT, PSL
- TIC,TCV, TI, TSH, TE, TY
- ZC, ZSH, ZSL ZT

Typical Letter Combinations

The table below gives a general definition of how letters are used in P&IDs.

	First letter		Succeeding letters		
	Measured or initiating variable	Modifier	Readout or passive function	Output function	Modifier
A	Analysis		Alarm		
B	Burner, combustion		User's choice	User's choice	User's choice
C	User's choice			Control	
D	User's choice	Differential			
E	Voltage		Sensor (primary element)		
F	Flow rate	Ration (fraction)			
G	User's choice		Glass, viewing device		
H	Hand				High
I	Current (electrical)		Indication		
J	Power	Scan			
K	Time, time schedule	Time rate of change		Control station	
L	Level		Light		Low
M	User's choice	Momentary			Middle, intermediate
N	User's choice		User's choice	User's choice	User's choice
O	User's choice		Orifice, restriction		
P	Pressure, vacuum		Point (test connection)		
Q	Quantity	Integrate, totalizer			
R	Radiation		Record		
S	Speed, frequency	Safety		Switch	
T	Temperature			Transmit	
U	Multivariable		Multifunction	Multifunction	Multifunction
V	Vibration, mechanical analysis			Valve, damper, louver	
W	Weight, force		Well		
X	Unclassified	X axis	Unclassified	Unclassified	Unclassified
Y	Event, state, or presence	Y axis		Relay, compute, convert	
Z	Position, dimension	Z axis		Driver, actuator	

Consider the following P&ID diagram:

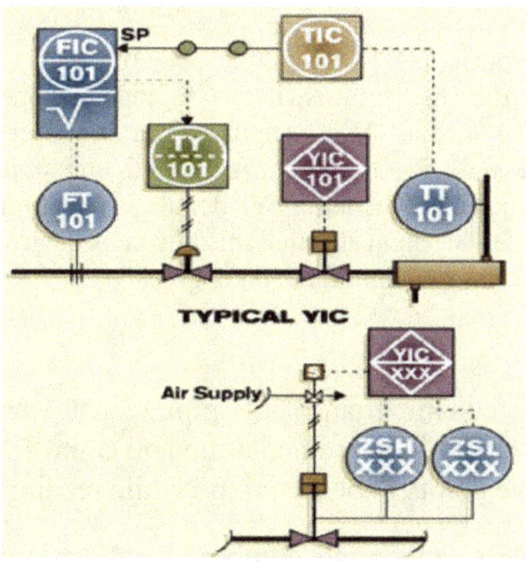

The flow indicating controller FIC 101, which is located in a shared control/display device, is electrically connected to the field-mounted flow transmitter FT 101 through a dotted line, as shown in the P&ID figure. Applying a square root extraction of the input signal is one of the features of FIC 101. The output of FIC 101 is an electrical signal to TY 101, which is located behind the panel board or in an inaccessible position. Because of its output signal—a pneumatic signal, which is a line with two forward slash marks—TY 101 is classified as an I/P (current to pneumatic transducer). TT 101 and TIC 101 measure, indicate, and control temperature similarly to FT 101 and FIC 101. The output of TIC 101 is connected via internal software or a data link (line with bubbles) to the set point (SP) of FIC 101 in order to establish a cascade control scheme.

A cover page providing definitions for common and standard terms, symbols, numbering schemes, etc. is often included with P&IDs. In this instance, the P&IDs would likely have the condensed form of

YIC, whereas the cover page would most likely contain the standard YIC.

The term "typical YIC" refers to a solenoid valve-controlling an on/off valve that has limit switches to indicate open (ZSH) and closed (ZSL) positions. Every input and output is connected to an operator-accessible PLC (shown as a solid horizontal line with a diamond in a square). The letter "Y" denotes a situation, existence, or event. The letter "C" indicates that this device is controlled, while the letter "I" indicates that an indication is given.

Exercise 8

1. Explain the importance of piping and instrumentation diagrams (P&IDs) in instrumentation and control.
2. Explain the points to bear in mind while reading P&IDs.

3. Explain how letters and numbers combinations are used in P&IDs.

4. List five safety devices necessary for the installation of instrument and control devices.

5. a) Enumerate the four graphical elements used in P&IDs.

b) List the three common location types different devices in P&IDs.

6. Explain the operating principle for each of the following industrial safety devices:

    i.   Pressure relief valve
    ii.   Rupture disc
    iii.   Flammable arrestor

7. Explain the reason why safety devices are used in industries.

8. Sketch the symbol for the following P&ID components:
i. tray column   ii. spiral heat exchanger   iii. Butterfly valve
iv. Angle valve   v. solenoid
9. Explain the principle behind the use of tag numbers and identification letters in P&ID diagrams.

10. Describe how to represent the following functions in P&IDs:

 i. discrete instruments
 ii. shared control/display
 iii. computer function
 iv. programmable logic controller

CHAPTER NINE

COMPUTER CONTROL SYSTEMS IN INDUSTRIES

9.1 Introduction

Since the information era began, computer control systems have been progressively used in many facets of life and industry. As technology has advanced, so too has the range of applications for these systems expanded. Computer-controlled systems are progressively replacing traditional methods of using human resources, particularly in the operation of industrial operations. The fundamental reason and purpose of the instrumentation systems has remained unaltered over a long time. However, there has been a paradigm shift in measurement, control, and instrumentation methodologies due to ongoing technological innovation and the emergence of faster and more accurate digital technologies and components like transducers, microprocessors, and analog-to-digital converters. These technologies have replaced manual monitoring and control.

Industrial process engineers and designers are increasingly looking to automate processes wherever technology allows. Thus they consistently brought in measuring instruments. As a result, the need for better instrumentation and automated control has become a key driver for better operation of industry processes.

The term **industrial automation** refers to a group of technologies that enable the autonomous running of industrial processes and machines without the need for human operators. These technologies include robotics and computer software. In addition to lowering expenses and saving time, industrial automation also provides better performance by removing the chance of human error.

Industrial automation requires a large variety of tools. These comprise a range of control systems that integrate diverse devices and systems that influence distinct facets of the production procedure. Following is a list of the main tools:

- Supervisory Control and Data Acquisition (SCADA)
- Human Machine Interface (HMI)
- Programmable Logic Controller (PLC)
- Artificial Neural Network (ANN)
- Distributed Control System (DCS)
- Robotics

9.2   Role of computers in measurement and control

The usage of computers for measurement and control applications has significantly risen as a result of the advancements in digital computer technology. Getting data from field devices (input) and computing a rational decision to change the material and energy flow of a process in a desired way to obtain the best possible output is the fundamental goal of computer-based measurement and control. In comparison to a general purpose computer, a process computer is expected to perform more rigidly and reliably in terms of response speed, computing power, adaptability, and fault tolerance. In addition, real-time process control is required. Other challenges faced by process computers mostly relate to finding a solution for the issues of complexity, adaptability, and geographic dispersion of process elements (plant equipment) that need to be operated under control.

Applications of digital computer control might be either passive or active in the process industries. While active application involves both the acquisition and manipulation of data and uses it for (real-time) process management, passive application just involves the capture of process data (data acquisition / data logging). The systems for data reduction, warning, and monitoring are the main focus of the passive application. Through an interface module, the process data is transmitted to the data acquisition computer after being obtained (measured) online. The operator can obtain real-time process measurement data and automatic transmission in the necessary format for further processing by the process control computer thanks to the smart instruments (smart sensors, smart transmitters, and smart actuators, the last control element) with embedded computers. The smart instruments guarantee that the actuator, transmitter, or sensor operates in accordance with the user's requirements.

Digital computers are mostly used in plant optimization and process control. Previously unaffordable, computer control systems may now be economically customized to suit the majority of industrial applications. Many people have been inspired by the advancements in computer control, which have also altered our understanding of how industrial processes operate. By using video display terminals, operators can now monitor the entire plant from a control room. A few keyboards and screens take the role of a large panel of instruments, knobs, and switches.

Nowadays, control rooms are significantly smaller, and plant supervision requires fewer personnel. Sophisticated mathematical models can now be implemented by process control computers. To support efficient operation, plant managers and engineers can receive thorough information about the state of plant operations. It is possible to automatically adjust controller parameters for optimal operating performance with the use of microprocessor-based devices and recently developed methodologies. Computer assistance is used to use expert systems and sophisticated control strategies, such as model-based predictive control, to optimize process performance.

9.3   Computer-based Data acquisition

A data acquisition system (DAQ) is a system that collects, stores, visualizes, and processes data. It consists of measurement equipment, sensors, and a computer, as well as data acquisition software. The purpose of a data acquisition system is to collect the information needed to understand an electrical or physical phenomenon. Data acquisition is the process of gathering information from real-world data sources. The goal is to analyze, process, and store the data in a computer. A common example of data acquisition is to measure a room's temperature as a digital value using sensor technology. In the industries, supervisory control and data acquisition (SCADA) systems are used along with the human machine interface (HMI) to carry our complex control of machinery.

9.3.1   SCADA

SCADA stands for Supervisory Control and Data Acquisition. Remote monitoring of activities and processes is made possible by this system, which regulates and gathers data on industrial output. In other words, SCADA refers to a technology that helps organisations and industries manage and track manufacturing production scheduling.

SCADA systems are essential for maintaining productivity in business decision-making. SCADA reduces downtime in the communication of problems due to supporting features that make data processing easier and more intelligent. SCADA is composed of either PLC (Programmable Logic Controller) or RTU (Remote Terminal Unit). Both of these units act as a communication link with HMIs, machines, terminals and sensors. The SCADA software then sends the data to a computer and begins to process, analyze, and display the data. This allows employees and operators to analyze information and take decisions.

The SCADA system contains the basic software and hardware needed to make it run smoothly. Here are the components of a SCADA system:

A central monitoring station: This is the point at which one or more central servers are linked to create a system platform, the central host computer server (CHCS). The system will communicate with the software programs and HMI (human-machine interface) at the central monitoring station.

An intermediate data collection station: A programmable logic controller (PLC) or remote terminal unit (RTU) is usually used for this. These facilitate communication between actuators, such as control equipment, switches, sensors, and actuators, etc.

A communication system: The equipment and industrial communication network are situated here. Together with field-level components like sensors, transducers, and measuring devices, the system will also comprise converters that manage the field-level data sent to the server.

9.3.2   HMI

The term "human machine interface" (HMI) is often used interchangeably with "dashboard" or "screen". Many line operators, managers, and supervisors in the industry rely on a human machine interface (HMI) to turn complicated data into valuable information. For instance, they use a HMI to monitor machinery and make sure it is operating correctly. Visual, easy-to-read displays provide context and meaning to near-real-time information such as tank level, pressure, vibration, motor, and valve status, and more.

A human machine interface (HMI) is the operator control panel that is connected to a PLC, RTU, or in some cases, directly to an IED. HMIs supplant physically actuated switches, dials, and other controls with graphical representations of the control and advanced controls to influence the process.

Today's high-performance machine HMI systems provide managers and supervisors with much more than just control processes. By leveraging historical and trending data, they provide new ways to enhance product quality and streamline systems.

9.4　Principle of data collection in SCADA system

In order to commence the data collection process, the RTUs first gather information from other connected devices. This process requires the same amount of time as internal scanning. Subsequently, the servers carefully scan the RTUs to gather both information and data. To gather data, the servers send a request signal to the RTUs, which subsequently signals other devices.

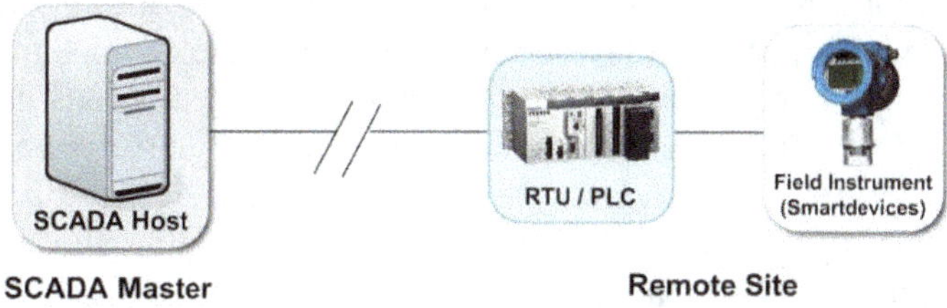

9.5　Industrial Automation Controllers

Three types of controllers are used in industrial automation: PLCs (programmable logic controllers), PACs (programmable automation controllers), and DCSs (distributed control systems). PLCs are one of these and they are important to the industries. An essential component of SCADA and HMI systems' hardware are PLCs. They give users the opportunity to view data from the manufacturing floor and access control and monitoring user interfaces. The actual links that connect SCADA or HMI systems to equipment on the factory floor or in plants are made by PLCs. PLCs are capable of monitoring, controlling, and interacting with a wide range of industrial gear, including temperature control, robot cells, conveyors, and many more.

PLCs are available in a wide range of shapes and sizes. While some are little enough to carry in your pocket, others are big enough to need to be mounted on their own sturdy racks. While some PLCs are more modular and simply have basic inputs and outputs (I/O), others can be tailored to match a variety of industrial applications by adding more back planes and functional modules (such analog I/O, communications modules, or display modules).

Programmable controllers and industrial computers are similar. PLCs can be used as standalone devices to automate and continuously watch over a certain process or machine operation. An complete production line can be controlled by networking PLCs. For almost any application, PLCs may be set up to process electrical signals and use them to execute preprogrammed commands. Numerous different types of sensors and actuators can be monitored and controlled by them. PLCs improve system performance, stability, and dependability in industrial automation by eliminating the need for human operators and the potential for human error..

PLCs are fast, simple to use, and easy to program, which makes them popular in many different industries. PLC programming is available in five mainstream languages. The most popular language is Ladder Logic, however Function Block Diagrams, Sequential Function Charts, Structured Text, and Instruction Lists can all be used to achieve the same goals.

Exercise 9

1. Write short notes on the following components of industrial automation:

    i.   Supervisory Control and Data Acquisition (SCADA)
    ii.   Human Machine Interface (HMI)

iii.   Programmable Logic Controller (PLC)
iv.   Artificial Neural Network (ANN)
v.   Distributed Control System (DCS)
vi.   Robotics

2. Explain briefly the term "industrial automation".

3. Explain the importance of PLC in industrial automation.

4. Explain briefly the following tools in industrial control system:

i.   SCADA
ii.   HMI
iii.   PLC
iv.   DCS
v.   Robotics